HARRAP'S

Chemistry

MINI DICTIONARY

Edited by
John O.E. Clark
and
William Hemsley

HARRAP
London

First published in Great Britain 1991
by HARRAP BOOKS Ltd
Chelsea House, 26 Market Square, Bromley
Kent BR1 1NA

ISBN 0 245–60325–5

Typeset by Action Typesetting, Gloucester
Printed in Great Britain by
Richard Clay Ltd,
Bungay, Suffolk

Preface

Harrap's *Chemistry Mini Dictionary* has been devised for two main groups of readers. The first group consists of people whose daily work brings them into contact with chemical terms. They may be industrial chemists or – because of the nature of their jobs – non-chemists who nevertheless have to understand specialist terminology. Readers in the second main category are students, and for them we have tried to meet several needs. Schoolchildren and students who are learning chemistry will find the Dictionary invaluable for checking the meanings of words that are a required part of thc vocabulary of the subject. Students in allied disciplines, such as biology, physics and medicine, can use it as a handy reference source for words from chemistry that are commonly employed in *their* subjects, but are so often taken for granted.

The overlap of chemistry with other disciplines posed a problem for the compilers of this Dictionary. There is a necessary progression from inorganic chemistry to organic chemistry, and then from these subdisciplines to biochemistry, and ultimately biology. So where does chemistry stop? Certainly the main terms of organic chemistry have to be included, as does some of the terminology of biochemistry – an organic chemist and a biochemist both have to know what a carbohydrate is (as does a medical student and a dietitian). Molecular biology continues to confirm the key role of chemistry in biological science. Yet we have taken the view that the 'living chemicals' that constitute the components of cells and their metabolism belong better to the *Biology Mini Dictionary*, a future companion volume to this one.

J.O.E.C. and W.T.J.H. – London, 1991

Introduction

Harrap's *Chemistry Mini Dictionary* provides an up-to-date reference source for a wide range of people. It contains definitions of many terms, and the entries are fully cross-referenced. The cross-references lead the reader both to definitions of unfamiliar terms and to further information about a subject. Cross-references are indicated by **bold type** within the text. For example, **'Ethanol'** in the article on absolute alcohol is printed in bold type; this indicates that ethanol has an article to itself in the Dictionary.

Over the last few years there has been a change in the way that the terminology of chemistry is presented. Some books use the older terminology; more recent publications employ the latest version. This dictionary lists names of chemicals in inorganic and organic chemistry in both their earlier and their more recent forms. No matter which term a reader looks up, the article will be found there, or else a cross-reference will be given to where it can be found. The name under which the main article appears has been selected to make the Dictionary easily accessible to the widest readership.

In order to help students using this book as a study aid, entries of particular relevance at GCSE level are indicated by ■ at the beginning of the entry. In addition, entries on topics that are part of Science in the National Curriculum are coded to show where they fit into the Curriculum. This code is at the end of the entry – for example [7/5/b]. The first number gives the attainment target, and the second number the level of attainment. The final letter indicates the statement of attainment within that level.

A

■ **absolute alcohol** **Ethanol** (ethyl alcohol) that contains no more than 1% water. [7/5/b]

absolute configuration Arrangement of groups about an asymmetric atom. *See* **configuration**.

■ **absolute zero** Lowest temperature theoretically possible, at which a substance has no heat energy whatever. It corresponds to −273.15°C, or zero on the kelvin scale. [13/3/b]

acetal $CH_3CH(OC_2H_5)_2$ Colourless volatile liquid organic compound. Alternative name: 1,1-diethoxyethane.

acetaldehyde CH_3CHO Colourless liquid organic compound with a pungent odour; a simple **aldehyde** made by the oxidation of **ethanol** (ethyl alcohol). Alternative names: acetic aldehyde, ethanal.

acetaldol Alternative name for **aldol**.

acetamide CH_3CONH_2 Colourless deliquescent crystalline organic compound, with a 'mousy' odour. Alternative name: ethanamide.

acetate *1*. Salt of **acetic acid** (ethanoic acid) in which the terminal hydrogen atom is substituted by a metal atom; *e.g.* copper acetate $Cu(CH_3COO)_2$. *2*. **Ester** of acetic acid in which the terminal hydrogen atom is substituted by a **radical**; *e.g.* ethyl acetate $CH_3COOC_2H_5$. Alternative name: ethanoate.

■ **acetic acid** CH_3COOH Colourless liquid **carboxylic acid**, with a pungent odour and acidic properties. It is the acid in vinegar. Alternative name: ethanoic acid.

acetone CH_3COCH_3 Colourless volatile liquid organic compound, with a sweetish odour; it is a simple **ketone**. Alternative name: 2-propanone.

acetonitrile CH_3CN Colourless liquid organic compound, with a pleasant odour. Alternative name: methyl cyanide.

acetophenone $C_6H_5COCH_3$ Colourless liquid organic compound, with a sweet pungent odour. Alternative name: phenyl methyl ketone.

acetyl group The group CH_3CO- (as in acetyl chloride). Alternative name: ethanoyl.

acetyl chloride CH_3COCl Colourless highly refractive liquid organic compound, with a strong odour. Alternative name: ethanoyl chloride.

■ **acetylene** C_2H_2 Colourless organic gas with an ether-like odour (when pure), used as a fuel and in organic synthesis. It is the simplest **alkyne** (olefin). Alternative name: ethyne.

acetylide Type of **carbide** resulting from the interaction of **acetylene** and a solution of a heavy metal salt. Most acetylides are unstable and explosive. Alternative name: ethynide.

■ **acid** Member of a class of chemical compounds whose aqueous solutions contain hydrogen ions. Solutions of acids have a **pH** of less than 7. Strong acids dissociate completely (into **ions**) in solution; weak acids only partly dissociate. An acid neutralizes a **base** to form a **salt**, and reacts with most metals to liberate hydrogen gas. [6/5/b]

■ **acid dye** Class of chemical colourants that are applied in weak **acid** solutions to wool, silk and polyamides (nylon).

■ **acidic oxide** Oxide of a non-metal, usually an **anhydride**, that reacts with water to form an **acid**, and with a **base** to form a

salt and water; *e.g.* sulphur dioxide, SO_2, reacts with water to form sulphurous acid, H_2SO_3. *See also* **basic oxide**. [7/5/c/]

■ **acid rain** Phenomenon caused by the pollution of the atmosphere with sulphur oxides and nitrogen oxides, which are produced largely by burning fossil fuels. The most common of these oxides are **sulphur trioxide** (SO_3), which combines readily with water to form **sulphuric acid** (H_2SO_4), and **nitrogen dioxide** (NO_2), which combines with water to form **nitric acid** (HNO_3). These acids are precipitated with snow and rain. [5/6/c]

■ **acid salt** Salt formed when not all the replaceable hydrogen atoms of an **acid** are substituted by a metal or its equivalent; *e.g.* sodium hydrogencarbonate (bicarbonate), $NaHCO_3$, diammonium hydrogenphosphate, $(NH_4)_2HPO_4$.

acid value Measure of the amount of free **fatty acid** present in fat or oil. The value is given as the number of milligrams of potassium hydroxide required to neutralize the fatty acids in 1g of the substance being tested.

acriflavine Orange crystalline organic compound used as an antiseptic. Alternative name: 2,8-diaminoacridine methochloride.

Acrilan Synthetic fibre produced by **copolymerization** of **acrylonitrile** and **vinyl acetate**.

acrolein $CH_2=CHCHO$ Colourless liquid organic compound, with a pungent odour. Alternative name: 2-propenal.

acrylic acid $CH_2=CH.COOH$ Reactive organic acid with a pungent odour, used to make **acrylic resins**. It is one of the olefin-monocarboxylic acids. Alternative name: propenoic acid.

acrylic resin Transparent **thermoplastic** formed by the

polymerization of **ester** or **amide** derivatives of **acrylic acid**. The resins are chiefly used in the manufacture of artificial fibres and for optical purposes, such as making lenses (*e.g.* Acrilan, Perspex).

acrylonitrile $H_2C=CHCN$ Colourless liquid **nitrile** used as a starting material in the manufacture of **acrylic resins**. Alternative names: vinyl cyanide, propenonitrile.

actinide Member of a series of **elements** in Group IIIB of the Periodic Table, of **atomic numbers** 90 to 103 (actinium, at. no. 89, is sometimes also included). All actinides are radioactive. Alternative name: actinoid.

actinium Ac Radioactive element in Group IIIB of the Periodic Table (usually regarded as one of the **actinides**); it has several **isotopes**, with half-lives of up to 21.7 years. It results from the decay of uranium–235. At. no. 89; r.a.m. 227.

■ **activated carbon** Charcoal treated so as to be a particularly good absorbent of gases.

■ **activation energy** Amount of energy required to initiate the breaking and re-formation of chemical **bonds**, and thus to start a chemical reaction. [7/10/a]

■ **active mass** Concentration of a substance that is involved in a chemical **reaction**. [8/9/a]

active site Part of an **enzyme** molecule to which its **substrate** is bound during **catalysis**.

acyclic Not cyclic; describing a chemical compound that does not contain a ring of atoms in its molecular structure.

acyl group Part of an organic compound that has the formula $RCO-$, where R is a **hydrocarbon** group (*e.g.* acetyl, CH_3CO-).

■ **addition polymer** **Polymer** formed from simple **monomers** by **addition reactions** (*e.g.* many kinds of *plastics*). [6/7/a]

■ **addition reaction** Chemical reaction in which one substance combines with another to form a third, without any other substance being produced. The term is most commonly used in organic chemistry; *e.g.* for the reaction between hydrogen bromide (HBr) and ethylene (ethyne, $CH_2=CH_2$) to form ethyl bromide (bromoethane, C_2H_5Br). *See also* **substitution reaction**.

adduct Product of the chemical combination of two atoms or molecules – *i.e.* of an **addition reaction**.

adipic acid $HOOC(CH_2)_4COOH$ Colourless crystalline organic compound, a major constituent of rosin, used in the manufacture of **nylon**. Alternative names: butanedicarboxylic acid, hexanedioic acid.

■ **adrenaline** **Hormone** secreted by the medulla (core) of the adrenal glands and at some nerve endings of the sympathetic nervous system. It is produced when the body prepares for violent physical action. Its effects include increased heartbeat, raised levels of sugar (glucose) in the blood and improved muscle action. Alternative names: adrenalin, epinephrine.

adsorbent Substance on which **adsorption** takes place.

adsorption Accumulation of **molecules** or **atoms** of a substance (usually a gas) on the surface of another substance (a solid or a liquid).

■ **aerobic** Describing a biochemical process that needs free oxygen in order to take place. *E.g.* aerobic respiration is the process by which cells obtain energy from the **oxidation** of fuel molecules by molecular oxygen with the formation of carbon dioxide and water. It yields more energy than anaerobic respiration (without free oxygen). [3/6/a]

■ **aerosol** *1*. Suspension of particles of a liquid or solid in a gas; a type of **colloid**. *2*. Device used to produce such a suspension.

■ **affinity** Tendency of some substances to combine chemically with others.

■ **agar** Complex **polysaccharide** obtained from seaweed, especially that of the genus *Gelidium*. It is commonly employed as a gelling agent in media used for growing micro-organisms and in food for human consumption. Alternative name: agar-agar.

■ **agent orange** Very poisonous mixture of two weedkillers used as a herbicide in warfare to destroy an enemy's crops and defoliate trees.

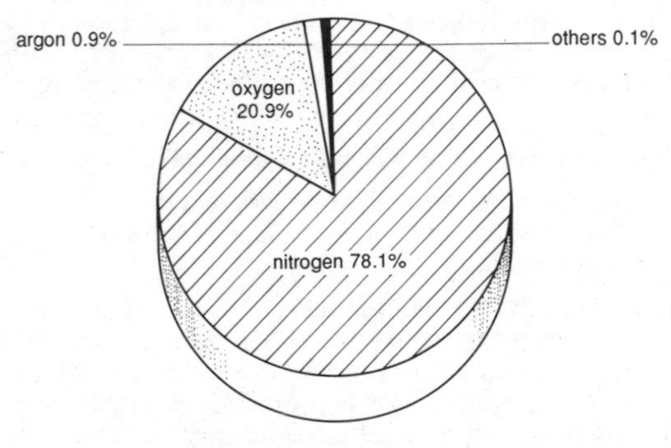

Average composition of air

■ **air** Mixture of gases that forms the Earth's **atmosphere**. Its composition varies slightly from place to place – particularly with regard to the amounts of carbon dioxide and water vapour it contains – but the average composition of dry air is (percentages by volume):

nitrogen 78.1%
oxygen 20.9%
argon 0.9%
other gases 0.1% [7/5/a]

-al Suffix usually denoting that an organic compound is an **aldehyde**; *e.g.* methanal (formaldehyde), ethanal (acetaldehyde).

■ **alabaster** Fine-grained compact **gypsum**.

alanine $CH_3C(NH_2)COOH$ **Amino acid** commonly found in proteins. Alternative name: 2-aminopropanoic acid.

■ **albumen** White of an egg. [3/8/a]

■ **albumin** Soluble **protein** found in many animal fluids, most notably in egg white and blood serum.

■ **alchemy** Predecessor of chemistry. Its two principal goals were the transmutation of common metals to gold, and the discovery of a universal remedy, the elixir.

■ **alcohol** *1.* Member of a large class of organic compounds that contain **hydroxyl (–OH) groups**. Simple, or primary, alcohols have the general formula ROH, where R is an **alkyl group** (or H in the case of methanol). Secondary alcohols have the formula RR′(CH)OH, and tertiary alcohols are RR′R″COH. An alcohol with two hydroxyl groups is called a **diol**, or glycol. Alcohols react with **acids** to form **esters**. A compound with a hydroxyl group attached to an **aryl group** is a **phenol**. *2.* Alternative name for **ethanol**. [7/5/b]

aldehyde Member of a large class of organic compounds that have the general formula RCHO, where R is an **alkyl** or **aryl group**. Aldehydes may be made by the controlled oxidation of **alcohols**. Their systematic names end with the suffix *-al* (*e.g.* the systematic name of acetaldehyde, CH_3CHO, is ethanal).

aldimine *See* **Schiff's base**.

aldol $CH_3CHOHCH_2CHO$ Viscous liquid organic compound. It is an **aldehyde**, a condensation product of **acetaldehyde** (ethanal). Alternative names: acetaldol, beta-hydroxy-butyraldehyde.

aldol reaction Condensation reaction between two **aldehyde** or two **ketone** molecules that produces a molecule containing an aldehyde (–CHO) group and an alcohol (–OH) group, hence *ald-ol* reaction.

aldose Type of **sugar** whose molecules contain an **aldehyde** (–CHO) group and one or more **alcohol** (–OH) groups.

alginic acid $(C_6H_8O_6)_n$ Yellowish-white organic solid, a polymer of mannuronic acid in the **pyranose** ring form, that occurs in brown seaweeds. Even very dilute solutions of the acid are extremely viscous, and because of this property it has many industrial applications.

alicyclic Describing a class of organic chemicals that possess properties of both **aliphatic** and **cyclic** compounds (*e.g.* cyclohexane, C_6H_{12}).

aliphatic Describing a large class of organic chemicals that have straight or branched chain arrangements of their constituent carbon atoms. The class includes **alkanes**, **alkenes**, **alkynes** and their derivatives.

alizarin $C_6H_4(CO)_2C_6H_2(OH)_2$ Orange-red crystalline organic compound important in the manufacture of dyes. Alternative name: 1,2-dihydroxy-anthraquinone.

■ **alkali** Substance that is either a soluble **base** or a solution of a base (*e.g.* sodium hydroxide, NaOH). Alkalis have a **pH** of more than 7, and react with acids to produce salts (and water). [6/5/b]

■ **alkali metal** One of the elements in Group I of the Periodic Table. They are **lithium, sodium, potassium, rubidium, caesium** and **francium**, which are all soft silvery metals that react vigorously with water. [7/6/a]

■ **alkaline** Having the properties of an **alkali**.

■ **alkaline earth** One of the elements in Group IIA of the Periodic Table. They are **beryllium, magnesium, calcium, strontium, barium** and **radium**. [6/9/a]

alkaloid One of a group of nitrogen-containing organic compounds that are found in some plants. Many are toxic or medicinal (*e.g.* atropine, digitalis, heroin, morphine, quinine, strychnine).

■ **alkane** Member of a group of saturated **aliphatic** hydrocarbons that have the general formula C_nH_{2n+2}; *e.g.* methane (CH_4), ethane (C_2H_6), etc. Alternative name: paraffin.

■ **alkene** Member of a group of unsaturated **aliphatic** hydrocarbons that have carbon-to-carbon double bonds and the general formula C_nH_{2n}; *e.g.* ethene (ethylene, C_2H_4), propene (propylene, C_3H_6), etc. Alternative name: olefin.

alkoxyalkane Alternative name for an **ether**.

alkyl group Hydrocarbon **radical**, derived from an **alkane** by the removal of one hydrogen atom, that has the general formula C_nH_{2n+1}; e.g. methyl ($-CH_3$), from methane, ethyl ($-C_2H_5$), from ethane, etc.

alkyl halide *See* **halogenoalkane**.

alkyl/aryl sulphide *See* **thioether**.

■ **alkyne** Member of a group of **unsaturated aliphatic** hydrocarbons that have carbon-to-carbon triple bonds and the general formula C_nH_{2n-2}; e.g. ethyne (acetylene, C_2H_2), propyne (methyl acetylene, C_3H_6), etc. Alternative name: acetylene.

■ **allotrope** One of the forms of an element that exhibits **allotropy**.

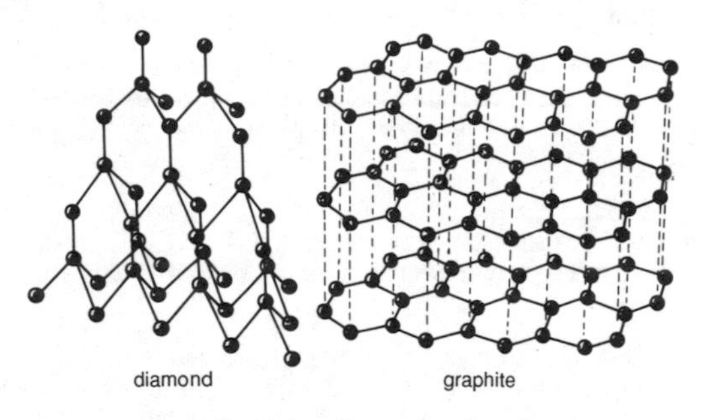

Two main allotropes of carbon

■ **allotropy** Existence of different structural forms of an element; *e.g.* graphite and diamond are allotropes of carbon. [8/8/c]

■ **alloy** Metallic substance that is made up of two or more elements (usually metals); *e.g.* brass (copper and zinc), bronze (copper and tin), solder (lead and tin) and steel (iron and carbon).

allyl group $CH_2=CHCH_2-$ **Unsaturated hydrocarbon radical** found in compounds such as allylbromide.

Alnico Trade name for an **alloy** that is composed mainly of aluminium, nickel and cobalt. It is used to make permanent magnets.

■ **alpha decay Radioactive** disintegration of a substance with the emission of **alpha particles**. [8/9/b]

alpha-iron Iron with a **body-centred cubic** structure.

■ **alpha particle** Particle that is composed of two **protons** and two **neutrons** (and is thus the equivalent of a helium nucleus). It is produced by **radioactive** decay, and has little penetrative power. [8/9/b]

■ **alum** *1.* $Al_2(SO_4)_3.K_2SO_4.24H_2O$ Aluminium potassium sulphate, a white crystalline substance (that occurs naturally as kalinite) used in leather-making and as a mordant in dyeing. *2.* Any of a group of salts with an analogous composition to alum (*i.e.* a double sulphate of a trivalent and monovalent metal with 24 molecules of **water of crystallization**).

■ **alumina** Another name for **aluminium oxide**.

aluminate Type of salt formed when **aluminium oxide** or hydroxide dissolves in strong alkali.

■ **aluminium** Al Silvery-white metallic element in Group IIIA of the Periodic Table. It occurs (as **aluminosilicates**) in many rocks and clays, and in **bauxite**, its principal ore, from which it is extracted by **electrolysis**. It is a light metal, protected against corrosion by a surface film of oxide. Its alloys are used in the aerospace industry and in light-weight structures. At. no. 13; r.a.m. 26.9815. [7/7/c]

aluminium chloride $AlCl_3$ White or yellowish deliquescent solid, covalently bonded when anhydrous, which fumes in moist air. It is used as a catalyst in the **cracking** of petroleum hydrocarbons.

■ **aluminium oxide** Al_2O_3 White crystalline **amphoteric** compound, the principal source of **aluminium**, which occurs as the mineral **corundum** and in **bauxite**. A thin surface film of aluminium oxide gives aluminium and its alloys their corrosion-resistant properties. Such films can be created by **anodizing**. Alternative name: alumina.

aluminium sulphate $Al_2(SO_4)_3$ White crystalline compound, used as a flocculating agent in water treatment and sewage works, as a **mordant** in dyeing, as a size in paper-making and as a foaming agent in fire extinguishers.

aluminium trimethyl Alternative name for **trimethylaluminium**.

aluminosilicate Common chemical compound in minerals and rocks (*e.g.* clays, mica) consisting of **alumina** and **silica** with water and various bases; such compounds are also formed in glass and various ceramics.

■ **amalgam** Mixture of mercury with one or more other metals, used for dental fillings.

amatol High explosive consisting of a mixture of **nitroglycerine** and **ammonium nitrate**.

americium Am Radioactive element in Group IIIB of the Periodic Table (one of the **actinides**). It has several **isotopes**, with half-lives of up to 7,650 years. It is used as a source of alpha-particles. At. no. 95; r.a.m. 243 (most stable isotope).

amide Member of a group of organic chemical compounds in which one or more of the hydrogen atoms of **ammonia** (NH_3) have been replaced by an **acyl group** ($-RCO$); *e.g.* acetamide (ethanamide), CH_3CONH_2. In primary amides one, in secondary amides two and in tertiary amides three of the hydrogens have been so replaced. Alternative name: alkanamides.

aminase One of a group of **enzymes** that can catalyse the **hydrolysis** of **amines.**

amination Transfer of an **amino group** ($-NH_2$) to a compound.

amine Member of a group of organic chemical substances in which one or more of the hydrogen atoms of **ammonia** (NH_3) have been replaced by a **hydrocarbon** group; *e.g.* methylamine, CH_3NH_2, aniline (aminobenzene or phenylamine), $C_6H_5NH_2$. In primary amines one, in secondary amines two and in tertiary amines three hydrogens have been so replaced.

■ **amino acid** The building blocks of **proteins,** amino acids are organic compounds that contain an acidic **carboxyl group** ($-COOH$) and a basic **amino group** ($-NH_2$). Twenty amino acids are commonly found in proteins. Those that can be synthesized by a particular organism are known as 'non-essential'; 'essential' amino acids must be obtained from the environment, usually from food. [3/7/b]

aminobenzene Alternative name for **aniline.**

aminoethanamide Alternative name for **guanidine.**

amino group Chemical group with the general formula $-NRR'$, where R and R′ may be **hydrogen** atoms or organic **radicals;** the commonest form is $-NH_2$. Compounds containing amino groups include **amines** and **amino acids**. *See also* **amide**.

aminoisovaleric acid Alternative name for **valine**.

aminoplastic resin Synthetic **resin** derived from the reaction of **urea, thiourea** or **melamine** with an **aldehyde**, particularly **formaldehyde** (methanal).

aminotoluene Alternative name for **toluidine**.

■ **ammonia** NH_3 Colourless pungent gas, which is very soluble in water (to form ammonium hydroxide, or ammonia solution, NH_4OH) and alcohol. It is formed naturally by the bacterial decomposition of proteins, purines and urea; made in the laboratory by the action of alkalis on ammonium salts; or synthesized commercially by fixation of nitrogen. Liquid ammonia is used as a refrigerant. The gas is the starting material for making nitric acid and nitrates. *See also* **Haber process**. [7/8/a]

■ **ammonia-soda process** Alternative name for **Solvay process**.

■ **ammonia solution** Alternative name for **ammonium hydroxide**.

ammonium carbonate $(NH_4)_2CO_3$ Unstable white crystalline compound which decomposes spontaneously to produce **ammonia**, used in smelling salts. The substance known in industry as ammonium carbonate is usually a double salt consisting of ammonium hydrogencarbonate (bicarbonate) and ammonium aminomethanoate (carbamate). Alternative name: sal volatile.

ammonium chloride NH_4Cl Colourless or white crystalline compound, used in dry batteries, as a **flux** in soldering and as a **mordant** in dyeing. Alternative name: sal ammoniac.

ammonium hydroxide NH_4OH **Alkali** made by dissolving **ammonia** in water, probably containing **hydrates** of ammonia. It is used for making soaps and fertilizers. 880 ammonia is a saturated aqueous solution of ammonia (density 0.88 g cm^{-3}). Alternative name: ammonia solution.

■ **ammonium ion** The ion NH_4^+, which behaves like a metal ion.

■ **ammonium nitrate** NH_4NO_3 Colourless crystalline compound, used as a fertilizer and in explosives.

ammonium phosphate $(NH_4)_3PO_4$ Colourless crystalline compound, used as a fertilizer (when it adds both nitrogen and phosphorus to the soil).

■ **ammonium sulphate** $(NH_4)_2SO_4$ White crystalline compound, much used as a fertilizer.

■ **amorphous** Without clear shape or structure.

■ **amphoteric** Describing a chemical compound with both basic and acidic properties. *E.g.* aluminium oxide, Al_2O_3, dissolves in acids to form aluminium salts and in alkalis to form aluminates.

■ **amu** Abbreviation of **atomic mass unit**.

■ **amylase** Member of a group of **enzymes** that digest **starch** or **glycogen** to **dextrin, maltose** and **glucose**. Amylases are present in digestive juices and micro-organisms. Alternative name: diastase. [3/7/b]

amyl group $C_5H_{11}-$ Monovalent **alkyl group**. Alternative name: pentyl group.

amyl nitrite $C_5H_{11}ONO$ Pale brown volatile liquid organic compound, often used in medicine to dilate the blood vessels of patients with some forms of heart disease (*e.g.* angina).

amylose Polysaccharide sugar, a polymer of **glucose** that occurs in **starch**.

amylum Alternative name for **starch**.

■ **anabolic steroid** Compound that is concerned with anabolism (the building-up of biochemicals). Commonly used anabolic steroids are synthetic male sex hormones (androgens) which promote protein synthesis (hence their use by some athletes wishing to build up muscle). *See also* **steroid**.

■ **anaerobic** Describing a biochemical reaction that takes place in the absence of free oxygen. *E.g.* anaerobic respiration is the process by which organisms obtain energy from the breakdown of food molecules in the absence of oxygen, as in fermentation. It yields less energy than **aerobic** respiration. [3/6/a]

analysis *See* **chromatography**; **qualitative analysis**; **quantitative analysis**; **thermal analysis**; **volumetric analysis**.

-ane Suffix usually denoting that an organic compound is an **alkane** or **cycloalkane**; *e.g.* methane, cyclohexane.

anhydride Chemical compound formed by removing water from another compound (usually an **acid**); *e.g.* an acid anhydride or **acidic oxide**.

anhydrite Naturally occurring **calcium sulphate**, used to make fertilizers. *See also* **gypsum**.

■ **anhydrous** Describing a substance that is devoid of moisture, or lacking **water of crystallization**.

aniline $C_6H_5NH_2$ Colourless oily liquid organic compound, one of the basic chemicals (feedstock) used in the manufacture of dyes, pharmaceuticals and plastics. Alternative names: aminobenzene, phenylamine.

animal starch Alternative name for **glycogen**.

■ **anion** Negatively charged **molecule** or **atom**.

■ **annealing** Process of bringing about a desirable change in the properties of a metal or glass (*e.g.* making it tougher), by heating it to a predetermined temperature, thus altering its microscopic structure.

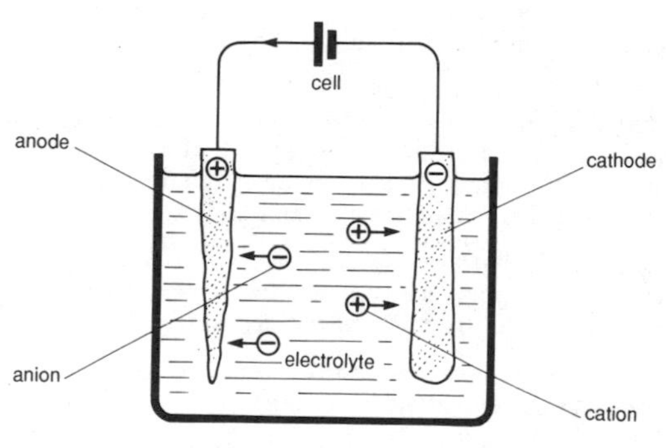

Anode is the positive electrode

■ **anode** Positive terminal of an **electrolytic cell** or thermionic valve.

■ **anodizing** Process of coating a metal (*e.g.* aluminium) with a thin layer of oxide by **electrolysis**.

anthracene $C_{14}H_{10}$ White solid aromatic compound whose molecules consist of three **benzene rings** fused in line.

■ **anthracite** Highest-grade coal with a very high carbon content (more than 88%, and up to 98%). It burns with a blue flame, and gives off little smoke.

anthraquinone $C_6H_4(CO)_2C_6H_4$ Yellow crystalline organic compound important in the manufacture of dyes.

■ **antimatter** Matter that is composed of **antiparticles**.

antimony Sb Blue-white semimetallic element in Group VA of the Periodic Table. It is used to impart hardness to lead-tin alloys and as a **donor** impurity in **semiconductors**. Its compounds, used as pigments, are poisonous. At. no. 51; r.a.m. 121.75.

■ **antioxidant** Substance used to delay the **oxidation** of paint, plastics and food by molecular oxygen. Most antioxidants are organic compounds; natural ones are found in vegetable oils and some fruits (as ascorbic acid). [7/7/b]

antiparticle Subatomic particle that corresponds to another particle of equal mass but opposite electric charge (*e.g.* a positron is the antiparticle of the electron).

■ **antiseptic** Substance that combats sepsis by killing or preventing the growth of bacteria.

apatite Mineral that consists mainly of **calcium phosphate**, used as a source of **phosphorus** and for making fertilizers.

aqua fortis Obsolete term for concentrated **nitric acid**.

aquamarine Blue form of the mineral **beryl**.

■ **aqua regia** Mixture of one part concentrated **nitric acid** to three parts concentrated **hydrochloric acid**, so called because it dissolves the 'noble' metals gold and platinum.

■ **aqueous** Dissolved in water, or chiefly consisting of water.

■ **aqueous solution** Solution in which the **solvent** is water.

arabinose $C_5H_{10}O_5$ Crystalline **pentose sugar** derived from plant **polysaccharides** (such as gums).

aragonite Fairly unstable mineral form of **calcium carbonate** ($CaCO_3$).

argentiferous Silver-bearing, usually applied to mineral deposits.

■ **argon** Ar Inert gas element in Group 0 of the Periodic Table (the **rare gases**). It makes up 0.9% of air (by volume), from which it is extracted. It is used to provide an inert atmosphere in electric lamps and discharge tubes, and for welding reactive metals (such as aluminium). At. no. 18; r.a.m. 39.948. [6/9/a]

aromatic compound Member of a large class of organic chemicals that exhibit **aromaticity**, the simplest of which is **benzene**.

aromaticity Presence in an organic chemical of five or more carbon atoms joined in a ring that exhibits **delocalization** of electrons, as in **benzene** and its compounds. All the carbon-carbon bonds are equivalent.

arsenate Salt or **ester** of **arsenic acid** (H_3AsO_4). Alternative name: arsenate(V).

arsenic As Silver-grey semimetallic element in Group VA of the Periodic Table which exists as several **allotropes**. It is used in alloys, as a **donor** impurity in **semiconductors** and in insecticides and drugs; its compounds are very poisonous. The substance known as white arsenic is arsenic(III) oxide (arsenious oxide), As_4O_6. At. no. 33; r.a.m. 74.9216.

arsenic acid H_3AsO_4 Tribasic acid from which **arsenates** are derived; an aqueous solution of arsenic(V) oxide, As_2O_5. Alternative names: orthoarsenic acid, arsenic(V) acid.

arsenide Compound formed from **arsenic** and another metal; *e.g.* iron(III) [ferric] arsenide, $FeAs_2$.

arsenious acid H_3AsO_3 Tribasic **acid** from which **arsenites** are derived; an aqueous solution of arsenic(III) oxide (arsenious oxide, white arsenic), As_4O_6. Alternative names: arsenic(III) acid, arsenous acid.

arsenite Salt or **ester** of **arsenious acid** (H_3AsO_3). Alternative name: arsenate(III).

arsine H_3As Colourless highly poisonous gas with an unpleasant odour. Organic derivatives, in which **alkyl groups** replace one or more hydrogen atoms, are also called arsines. Alternative name: arsenic(III) hydride.

■ **artificial radioactivity** Radioactivity in a substance that is not normally radioactive. It is created by bombarding the substance with ionizing **radiation**. Alternative name: induced radioactivity. [8/9/c]

aryl group Radical that is derived from an **aromatic compound** by the removal of one hydrogen atom; *e.g.* phenyl, C_6H_5-, derived from **benzene**.

■ **asbestos** Fibrous variety of a number of rock-forming **silicate** minerals that are heat-resistant and chemically inert.

■ **ascorbic acid** $C_6H_8O_6$ White crystalline water-soluble **vitamin** found in many plant materials, particularly fresh fruit and vegetables. It is a natural **antioxidant**. Alternative name: vitamin C.

■ **aseptic** Free from disease-causing micro-organisms (particularly bacteria). [2/4/b]

■ **aspirator** Apparatus that produces suction in order to draw a gas or liquid from a vessel or cavity.

■ **aspirin** $CH_3COO.C_6H_4COOH$ Drug that is commonly used as an analgesic, antipyretic and anti-inflammatory. Alternative name: acetyl salicylic acid. [3/2/b] [3/6/c]

astatine At Radioactive element in Group VIIA of the Periodic Table (the **halogens**). It has several **isotopes**, of half-lives of 2×10^{-6} sec to 8 hr. Because of their short half-lives, they are available in only minute quantities. At. no. 85; r.a.m. (most stable isotope) 210.

■ **atmosphere** Air or gases surrounding the Earth or other heavenly body. The Earth's atmosphere extends outwards several thousand kilometres, becoming increasingly rarefied until it merges gradually into space. For composition, *see* **air**. [9/3/b]

■ **atom** Fundamental particle that is the basic unit of matter. An atom consists of a positively-charged **nucleus** surrounded by negatively-charged **electrons** restricted to **orbitals** of a given energy level. Most of the **mass** of an atom is in the nucleus, which is composed principally of **protons** (positively charged) and **neutrons** (electrically neutral); hydrogen is exceptional in having merely one proton in its nucleus. The number of electrons is equal to the number of protons, and this is the **atomic number**. The chemical behaviour of an atom is determined by how many electrons it has, and how they are transferred to, or shared with, other atoms to form **chemical bonds**. *See also* **atomic mass**; **Bohr theory**; **isotope**; **subatomic particle**; **valence**. [8/8/a]

■ **atomic bomb** Explosive device of great power that derives its energy from nuclear fission. Alternative names: atom bomb, nuclear bomb.

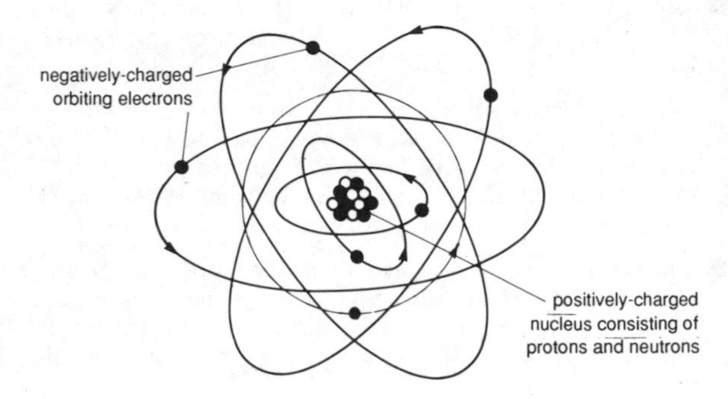

Bohr's model of a carbon atom

- **atomic energy** Energy that is released by **nuclear fission** or **nuclear fusion**. Alternative name: **nuclear energy**. [8/8/b]
- **atomicity** Number of **atoms** in one molecule of an element; *e.g.* argon has an atomicity of 1, nitrogen 2, ozone 3.
- **atomic mass** Alternative name for **atomic weight**. *See also* **relative atomic mass**.
- **atomic mass unit** (amu) Arbitrary unit that is used to express the mass of individual atoms. The standard is a mass equal to 1/12 of the mass of a carbon atom (the carbon – 12 **isotope**). A mass expressed on this standard is called a relative atomic mass (r.a.m.), symbol A_r. An alternative name for atomic mass unit is dalton. [8/8/a]

■ **atomic number** (at. no.) Number equal to the number of **protons** in the nucleus of an **atom** of a particular element, symbol Z. Alternative name: proton number. [8/8/d]

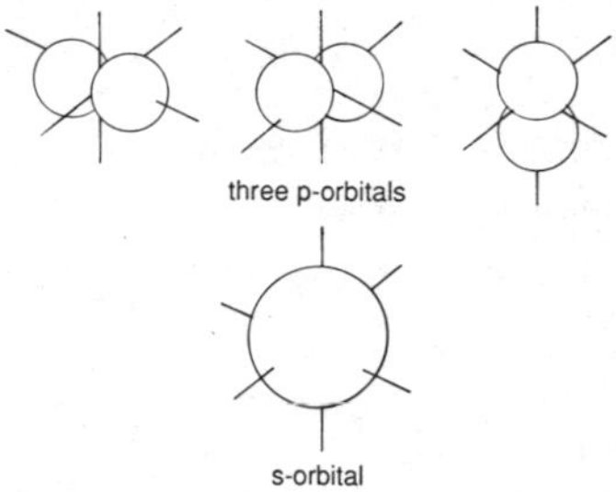

three p-orbitals

s-orbital

Atomic orbitals

atomic orbital Wave function that characterizes the behaviour of an **electron** when orbiting the nucleus of an atom. It can be visualized as the region in space occupied by the electron.

■ **atomic weight** (at. wt.) Relative mass of an **atom** given in terms of **atomic mass units**. *See* **Relative atomic mass**.

Auger effect Radiationless ejection of an **electron** from an **ion**. It was named after the French physicist Pierre Auger (1899–). Alternative name: **autoionization**.

auric Trivalent gold. Alternative names: gold(III), gold(3+).

auriferous Gold-bearing, usually applied to mineral deposits.

aurous Monovalent gold. Alternative names: gold(I), gold(1 +).

autocatalysis Catalytic reaction that is started by the products of a reaction that was itself catalytic. *See* **catalyst**.

autoclave Airtight container that heats and sometimes agitates its contents under high-pressure steam. Autoclaves are usually used for sterilization or industrial processing.

autoionization Spontaneous ionization of excited **atoms**, **molecules** or fragments of an **ion**. Alternative name: pre-ionization. *See also* **Auger effect**.

auto-oxidation **Oxidation** caused by the unaided **atmosphere**.

■ **Avogadro constant** (L) Number of particles (atoms or molecules) in a **mole** of a substance. It has a value of 6.02×10^{23}. It was named after the Italian scientist Amedeo Avogadro (1776–1856). Alternative name: Avogadro's number. [8/9/a]

■ **Avogadro's law** Under the same conditions of pressure and temperature, equal volumes of all gases contain equal numbers of **molecules**. Alternative name: Avogadro's hypothesis. [6/8/a]

■ **Avogadro's number** Alternative name for **Avogadro constant**.

avoirdupois System of weights, still used for some purposes in Britain, based on a pound (symbol lb), equivalent to 2.205kg, and subdivided into 16 ounces, or 7,000 grains. In science and medicine it has been almost entirely replaced by **SI units**.

azeotrope Mixture of two liquids that boil at the same temperature.

azide One of the **acyl group** compounds, or salts, derived from hydrazoic acid (N_3H). Most azides are unstable, and heavy metal azides are explosive.

azo dye Member of a class of dyes that are derived from amino compounds and have the $-N=N-$ **chromophore** group. They are intensely coloured (usually red, brown or yellow) and account for a large proportion of the synthetic dyes produced. Azo dyes can be made as acid, basic, direct or mordant dyes. Their use as food colourants has been questioned because of their possible effects on sensitive people, particularly children.

azomethine *See* **Schiff's base**.

B

Babbitt metal Tin alloy (containing also some antimony and copper) used for lining bearings in machinery. It was named after the American engineer Isaac Babbitt (1799–1862).

Babo's law The lowering of the **vapour pressure** of a **solution** is proportional to the amount of **solute**. It was named after the German chemist Lambert Babo (1818–99).

■ **background radiation Radiation** from natural sources, including outer space (cosmic radiation) and radioactive substances on Earth (*e.g.* in igneous rocks such as granite). [8/7/d]

bactericide Substance that can kill bacteria. *See also* **bacteriostatic**.

bacteriostatic Substance that inhibits the growth of bacteria without killing them. *See also* **bactericide**.

■ **Bakelite** Trade name for a **thermosetting** synthetic plastic made by the **condensation** of **phenol** with **formaldehyde** (methanal). It is used in the manufacture of electrical fittings and other plastic products. [7/9/b]

■ **baking powder** Mixture of a **hydrogencarbonate** (bicarbonate) and a weak **acid** which on heating or the addition of water produces bubbles of carbon dioxide; the bubbles cause a cake mixture or dough to rise. A common composition is a mixture of sodium hydrogencarbonate (baking soda) and tartaric acid.

■ **baking soda** Alternative name for **sodium hydrogencarbonate** (sodium bicarbonate).

Balmer series Visible atomic spectrum of hydrogen, consisting of a unique series of energy emission levels which appears as lines of red, blue and blue-violet light. It is the key to the discrete energy levels of electrons (*see* **Bohr theory**).

barium Ba Silver-white metallic element in Group IIA of the Periodic Table (the **alkaline earths**), obtained mainly from the mineral barytes (**barium sulphate**). Its soluble compounds are poisonous, and used in fireworks; its insoluble compounds are used in pigments and medicine. At. no. 56; r.a.m. 137.34.

■ **barium sulphate** $BaSO_4$ White crystalline insoluble powder, used as a pigment and as the basis of 'barium meal' to show up structures in X-ray diagnosis (because it is opaque to X-rays).

baryon One of a group of subatomic particles that include **protons**, **neutrons** and **hyperons**, and which are involved in strong interactions with other particles.

basalt Dark-coloured, usually black, fine-grained igneous rock. Deposits that occur on the Earth's surface are generally restricted to solidified lava.

■ **base** Member of a class of chemical compounds whose aqueous solutions contain OH^- **ions** (*e.g.* potassium hydroxide). A base neutralizes an **acid** to form a **salt** (*see also* **alkali**). [7/6/a]

■ **base metals** Metals that corrode, oxidize or tarnish on exposure to air, moisture or heat; *e.g.* copper, iron, lead.

■ **basic** In chemistry, having a tendency to release **hydroxide** (OH^-) ions. *See* **base**.

■ **basic oxide** Metallic oxide that reacts with water to form a **base**, and with an **acid** to form a **salt** and water; *e.g.* calcium oxide, CaO, reacts with water to form calcium hydroxide, $Ca(OH)_2$. *See also* **acidic oxide**. [7/5/c]

■ **basic salt** Type of **salt** that contains **hydroxide** (OH^-) ions; *e.g.* basic lead carbonate, $2PbCO_3.Pb(OH)_2$.

■ **battery** Device for producing electricity (direct current) by chemical action. Alternative name: cell. *See* **Daniell cell; dry cell; Leclanché cell; primary cell; secondary cell.** [11/3/b] [11/6/a]

Baumé scale Scale of relative density (specific gravity) of liquids, commonly used in continental Europe. It was named after the French chemist Antoine Baumé (1728–1804).

■ **bauxite** Earthy mineral form of **alumina** (aluminium oxide, Al_2O_3) and the chief ore of **aluminium.**

Beckmann thermometer Mercury thermometer used for accurately measuring very small changes or differences in temperature. The scale usually covers only 6 or 7 degrees. It was named after the German chemist Ernst Beckmann (1853–1923).

■ **becquerel** (Bq) SI unit of radioactivity, equal to the number of **atoms** of a **radioactive** substance that disintegrate in one second. It was named after the French physicist Henri Becquerel (1852–1908). 1 Bq = 2.7×10^{-11} curies (the former unit of radioactivity). [8/7/d]

Beer's law Concerned with the absorption of light by substances, it states that the fraction of incident light absorbed by a solution at a given wavelength is related to the thickness of the absorbing layer and the concentration of the absorbing substance. Alternative name: Beer-Lambert law.

beet sugar Alternative name for **sucrose.**

Benedict's test Food test used to detect the presence of a **reducing sugar** by the addition of a solution containing sodium carbonate, sodium citrate, potassium thiocyanate, copper sulphate and potassium ferrocyanide. A change in colour from blue to red or yellow on boiling indicates a positive result. It was named after the American chemist S. Benedict (1884–1936). *See also* **Fehling's test.**

bentonite Kind of clay, used as an **adsorbent** and in paper-making.

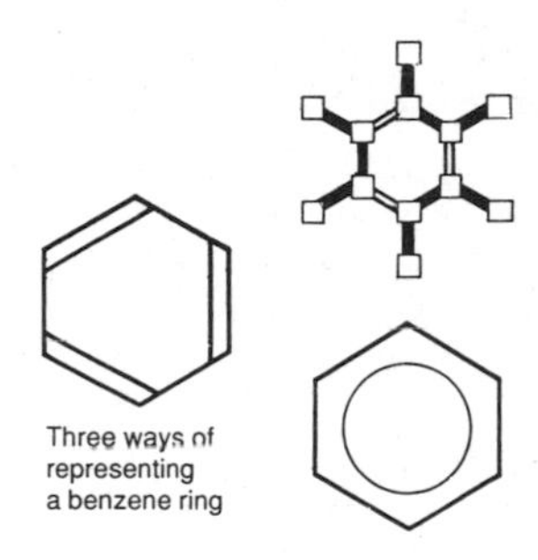

Various representations of the benzene ring

benzene C_6H_6 Colourless inflammable liquid **hydrocarbon,** the simplest **aromatic compound**. It is used as a solvent and in the manufacture of plastics.

1,3-benzenediol Alternative name for **resorcinol**.

benzene ring Cyclic (closed-chain) arrangement of six carbon atoms, as in a molecule of **benzene**. Molecules containing one or more benzene rings display **aromaticity**.

benzoic acid C_6H_5COOH White crystalline organic compound, used as a food preservative because it inhibits the growth of yeasts and moulds.

benzole Alternative name for **benzene**.

benzpyrene Cyclic organic compound, found in coal-tar and tobacco smoke, which has strong carcinogenic properties.

benzpyrrole Alternative name **indole**.

berkelium Bk Radioactive element in Group IIIB of the Periodic Table (one of the **actinides**). It has several **isotopes**, made by alpha-particle bombardment of americium-241. At. no. 97; r.a.m. (most stable isotope) 247.

beryl Beryllium aluminium silicate, a mineral which when not of gem quality is used as a source of **beryllium**.

beryllium Be Silver-grey metallic element in Group IIA of the Periodic Table (the **alkaline earths**). It is used for windows in X-ray tubes and as a **moderator** in nuclear reactors. At. no. 4; r.a.m. 9.0122.

■ **Bessemer process** Method of making steel which involves blowing air through molten iron to oxidize excess carbon and other impurities. It was named after the British engineer Henry Bessemer (1813–98). [7/7/c]

■ **beta decay** Disintegration of an unstable **radioactive** nucleus that involves the emission of a **beta particle**. It occurs when a neutron emits an electron and is itself converted to a proton, resulting in an increase of one proton in the nucleus concerned and a corresponding decrease of one neutron. This leads to the formation of a different element (*e.g.* beta decay of carbon-14 produces nitrogen). [8/9/c]

■ **beta particle** High-velocity electron emitted by a **radioactive** nucleus undergoing **beta decay**. [8/9/b]

■ **beta radiation** Radiation, consisting of beta particles (electrons), due to **beta decay**. [8/9/b]

■ **bicarbonate** Alternative name for **hydrogencarbonate**.

bichromate Alternative name for **dichromate**.

binary compound Chemical compound whose molecules consist of two different atoms.

binding energy Energy required to cause a **nucleus** to decompose into its constituent **neutrons** and **protons**.

biochemical oxygen demand (BOD) Oxygen-consuming property of natural water because of the organisms that live in it.

■ **biochemistry** Study of the **chemistry** of living organisms.

■ **biodegradable** Describing a substance that breaks down or decays by the action of living organisms, especially bacteria and fungi. Through biodegradation, organic matter is recycled. [2/4/b] [2/6/b]

■ **biotechnology** Utilization of living organisms for the production of useful chemical substances or processes, *e.g.* in **fermentation** and milk production. [7/5/b]

biotin Coenzyme that is involved in the transfer of carbonyl groups in biochemical reactions, such as the metabolism of fats; one of the B vitamins.

Birkeland-Eyde process Method for fixation of **nitrogen** in which air is passed over an electric arc to form **nitrogen monoxide** (nitric oxide). It was named after the Norwegian chemists Kristian Birkeland (1867–1913) and Samuel Eyde (1866–1940).

bismuth Bi Silvery-white metallic element in Group VA of the Periodic Table. It is used as a liquid metal coolant in nuclear reactors and as a component of low-melting point lead alloys. Its soluble compounds are poisonous; its insoluble ones are used in medicine. At. no. 83; r.a.m. 208.98.

■ **bisulphate** Alternative name for **hydrogensulphate**.

■ **bisulphite** Alternative name for **hydrogensulphite**.

■ **bitumen** Solid or tarry mixture of **hydrocarbons** obtained from coal, oil, etc., commonly used for surfacing paths and roads. [7/9/a]

■ **bituminous coal** Second-quality coal, with a carbon content above 65% but also containing large quantities of gas, water and coal tar. It is the most widely used domestic and industrial coal.

biuret $NH_2CONHCONH_2$ Colourless crystalline organic compound made by heating **urea**.

biuret test Test used to detect **peptides** and **proteins** in solution by treatment of **biuret** with **copper sulphate** and **alkali** to give a purple colour.

■ **bivalent** Having a **valence** of two. Alternative name: divalent.

■ **blast furnace** Furnace that can reach a high operating temperature, necessary for extracting iron from iron ore. The furnace is loaded with iron ore, coke and limestone. Hot air is blown into the mixture. Molten **pig iron** is run off from the bottom of the furnace. [7/7/c]

blasting glycerin Alternative name for **gelignite**.

■ **bleach** Substance used for removing colour from, *e.g.*, cloth, paper and straw. A common bleach is a solution of sodium chlorate(I) (sodium hypochlorite), NaClO, although hydrogen peroxide, sulphur dioxide, chlorine, oxygen and even sunlight are also used as bleaches. All are **oxidizing agents**.

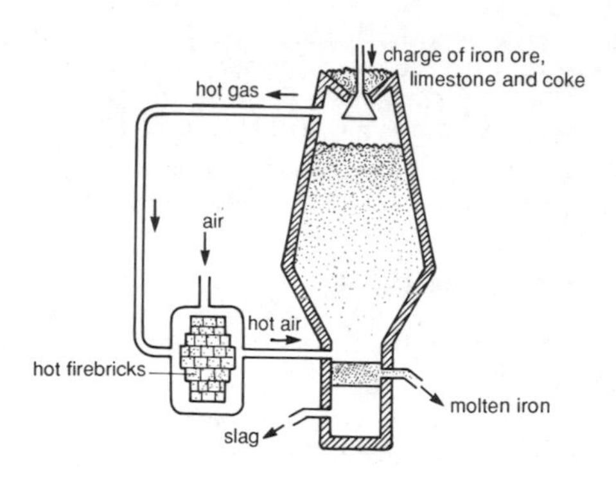

Blast furnace and its air supply

■ **bleaching powder** White powder containing calcium chlorate(I) (calcium hypochlorite), $Ca(OCl)_2$, made by the action of chlorine on calcium hydroxide (slaked lime). When treated with dilute acid it generates chlorine, which acts as a **bleach**.

blue vitriol Alternative name for **copper(II) sulphate**.

body-centred cube Crystal structure that is cubic, with an **atom** at the centre of each cube. Each atom is surrounded by eight others.

Bohr theory Atomic theory that assumes all **atoms** are made up of a central positively charged **nucleus** surrounded by orbiting planetary **electrons**, and which incorporates a **quantum theory** to limit the number of allowed **orbitals** in which the electrons can move. Each orbital has a characteristic energy level, and emission of electromagnetic radiation (*e.g.* light) occurs when an electron jumps to an orbital at a lower energy level (*see* **Balmer series**). It was named after the Danish physicist Niels Bohr (1885–1962).

■ **boiling point** (b.p.) Temperature at which a liquid freely turns into a vapour; the vapour pressure of the liquid then equals the external pressure on it.

Boltzmann constant (*k*) Equal to R/L = 1.3806×10^{-23} J K^{-1} (joule per kelvin), where R = gas constant and L = **Avogadro constant**. It was named after the Austrian physicist Ludwig Boltzmann (1844–1906).

■ **bond** Link between two **atoms** in a **molecule**. *See* **co-ordinate bond; covalent bond; ionic bond**. [7/10/a]

■ **bond energy** Energy involved in **bond** formation. [7/10/a]

■ **bond length** Distance between the **nuclei** of two **atoms** that are chemically bonded. [7/10/a]

boracic acid Alternative name for **boric acid**.

borane Any of the boron hydrides, general formula B_nH_{n+2}, which have unusual chemical bonding (with too few electrons for normal covalent bonds).

borax $Na_2B_4O_7.10H_2O$ White amorphous compound, soluble in water, which occurs naturally as **tincal**. It is used in the manufacture of enamels and heat-resistant glass. Alternative names: disodium tetraborate, sodium borate.

boric acid H_3BO_3 White crystalline compound, soluble in water, which occurs naturally in volcanic regions of Italy. It has **antiseptic** properties. Alternative name: boracic acid.

boron B Amorphous, non-metallic element in Group IIIA of the Periodic Table. Because of its high neutron absorption it is used for **control rods** in nuclear reactors. Important compounds include **borax** and **boric acid**. At. no. 5; r.a.m. 10.81.

■ **borosilicate glass** Heat-resistant glass of low thermal expansion, made by adding boron oxide (B_2O_3) to **glass** during manufacture.

■ **Bosch process** Industrial process for manufacturing **hydrogen** by the catalytic **reduction** of steam with **carbon monoxide**. It is used to produce hydrogen for the **Haber process**. It was named after the German chemist Carl Bosch (1874–1940). [7/7/b]

boson Subatomic particle, *e.g.* **alpha particle**, photon, that obeys Bose-Einstein statistics but does not obey the **Pauli exclusion principle**. Atomic **nuclei** of even mass numbers are also bosons.

■ **Boyle's law** At constant temperature the volume of a gas is inversely proportional to its pressure. It was named after the Irish chemist Robert Boyle (1627–91). [6/8/a]

■ **branched chain** Side group(s) attached to the main chain in the molecule of an organic compound.

■ **brass** Alloy of **copper** and **zinc**.

■ **breeder reactor Nuclear reactor** that produces more fissile material (*e.g.* plutonium) than it consumes. [13/8/c]

bromide Binary compound containing **bromine**; a salt of hydrobromic acid.

bromide paper Photographic (light-sensitive) paper coated with an emulsion containing silver bromide, used for making black-and-white prints and enlargements.

■ **bromine** Br Dark red liquid non-metallic element in Group VIIA of the Periodic Table (the **halogens**), extracted from sea-water. It has a pungent smell. Its compounds are used in photography and as anti-knock additives to petrol. At. no. 35; r.a.m. 79.904. [6/9/a]

■ **bromine water** Hypobromous acid (HBrO), made by dissolving **bromine** in water.

bromoform Alternative name for **tribromomethane**.

Brönsted-Lowry theory Concept of **acids** and **bases** in which an acid is defined as a substance with a tendency to lose a proton (H^+) and a base as a substance with a tendency to gain an electron. It was named after the British chemist Thomas Lowry (1874–1936) and the Danish chemist Johannes Brönsted (1879–1947), who proposed it independently. Alternative name: Lowry-Brönsted theory. *See also* **Lewis acid and base**.

■ **bronze** Alloy of **copper** and **tin**.

brown coal Alternative name for **lignite**.

■ **Brownian movement** Random motion of particles of a **solid** suspended in a **liquid** or **gas**, caused by collisions with molecules of the suspending medium. It was named after the British botanist Robert Brown (1773–1858). Alternative name: Brownian motion. [8/7/a]

■ **brown ring test** Laboratory test for **nitrates** in solution. An acidic solution of iron(II) sulphate (ferrous sulphate) is added to the nitrate solution in a test-tube, and concentrated sulphuric acid carefully poured down the inside of the tube. A brown ring at the junction of the liquids indicates the presence of a nitrate.

■ **buffer** Solution that resists changes in **pH** on dilution or on the addition of **acid** or **alkali**.

buna-S Alternative name for **styrene-butadiene rubber.**

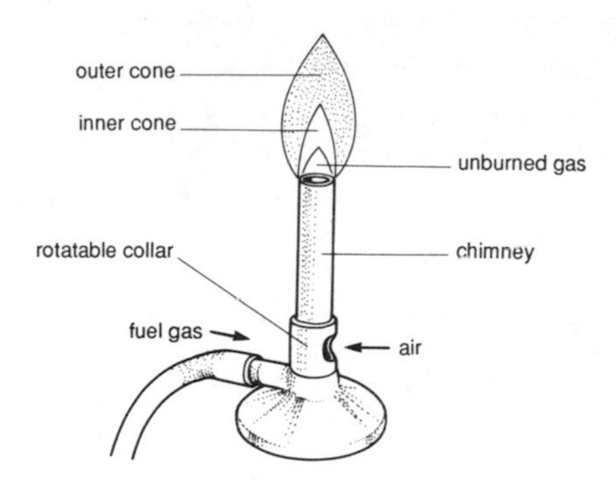

Bunsen burner

■ **Bunsen burner** Gas burner that efficiently mixes air with the fuel gas, commonly used for heating in laboratories. It was named after the German chemist Robert Bunsen (1811–99).

■ **burette** Long vertical graduated glass tube with a tap, used for the addition of controlled and measurable volumes of liquids (*e.g.* in making **titrations** in **volumetric analysis**).

butadiene $CH_2=CHCH=CH_2$ **Hydrocarbon** gas with two double bonds, important in making **polymers** such as synthetic rubber.

■ **butane** C_4H_{10} Gaseous **alkane**, used as a portable supply of fuel.

butanedicarboxylic acid Alternative name for **adipic acid**.

butanedioic acid Alternative name for **succinic acid**.

butanol C_4H_9OH Colourless liquid **alcohol** that exists in four isomeric forms. Alternative name: butyl alcohol.

butyl alcohol Alternative name for **butanol**.

butyl rubber Polymer produced by the polymerization of isobutene with small amounts of **isoprene**.

■ **by-product** Incidental or secondary product of manufacture.

C

cadmium Cd Silvery-white metallic element in Group IIB of the Periodic Table. It is used in **control rods** for nuclear reactors and as a corrosion-resistant electroplating on steel articles. Its compounds are used as yellow or red pigments. At. no. 48; r.a.m. 112.40.

caesium Cs Soft, reactive metallic element in Group IA of the Periodic Table (the **alkali metals**), a major fission product of uranium. It is used in photoelectric cells. The isotope caesium-137 is used in radiotherapy. At. no. 55; r.a.m. 132.905.

■ **caffeine** White alkaloid with a bitter taste, obtained from coffee beans, tea leaves, kola nuts or by chemical synthesis. It is a diuretic and a stimulant to the central nervous system.

■ **calamine** Zinc ore whose main constituent is zinc oxide.

■ **calcite** Crystalline form of natural **calcium carbonate**. It is a major constituent of limestones and marbles, and as such is one of the most common minerals.

■ **calcium** Ca Silver-white metallic element in Group IIA of the Periodic Table (the **alkaline earths**). The fifth most abundant element on Earth, it occurs mainly in **calcium carbonate** minerals; it also occurs in bones and teeth. At. no. 20; r.a.m. 40.08. [6/9/a]

■ **calcium carbide** CaC_2 Compound of carbon and calcium made commercially by heating coal and lime to a high temperature. It reacts with water to produce **acetylene**, and was once used for this purpose in acetylene lamps. Alternative names: calcium acetylide, calcium ethynide, carbide.

- **calcium carbonate** $CaCO_3$ White powder or colourless crystals, the main constituent of chalk, limestone and marble.
- **calcium chloride** $CaCl_2$ White crystalline compound, which forms several **hydrates**, used to control dust, as a de-icing agent and as a refrigerant. The anhydrous salt is **deliquescent** and is employed as a **desiccant**.
- **calcium hydrogencarbonate** $CaHCO_3$ White crystalline compound, stable only in solution and the cause of temporary **hardness of water**. Alternative name: calcium bicarbonate.
- **calcium hydroxide** $Ca(OH)_2$ White crystalline powder which gives an alkaline aqueous solution known as limewater, used as a test for carbon dioxide (which turns it cloudy). Alternative names: calcium hydrate, hydrated lime, caustic lime, slaked lime.
- **calcium oxide** CaO White crystalline powder made commercially by roasting **limestone** (calcium carbonate). It is used to make **calcium hydroxide** (slaked lime), for treating acid soils, for making mortar and in smelting iron and other metals (to help to form **slag**). Alternative names: lime, quicklime.
- **calcium phosphate** $Ca_3(PO_4)_2$ White crystalline solid which makes up the mineral component of bones and teeth, and occurs as the mineral **apatite**. It is produced commercially as bone ash and basic slag. Treated with sulphuric acid it forms the fertilizer known as superphosphate.
- **calcium silicate** Ca_2SiO_4 White insoluble crystalline compound, present in various minerals and cements, and a component in the slag produced in a **blast furnace**.
- **calcium sulphate** $CaSO_4$ White crystalline compound which occurs as the minerals anhydrite and (as the dihydrate)

gypsum, used to make **plaster of Paris**. It is the cause of permanent **hardness of water**. It is used in making ceramics, paint, paper and sulphuric acid, and is the substance in blackboard 'chalk'.

californium Cf Radioactive element in Group IIIB of the Periodic Table (one of the **actinides**), produced by alpha-particle bombardment of curium-242; it has several **isotopes**, with half-lives of up to 800 years. At. no. 98; r.a.m. 251 (most stable isotope).

calomel Alternative name for **mercury(I) chloride**.

calomel half-cell Reference **electrode** of known potential consisting of mercury, mercury(I) chloride and potassium chloride solution. Alternative names: calomel electrode, calomel reference electrode.

■ **Calor gas** Portable fuel gas consisting mainly of **butane** and **propane**, stored under pressure in metal bottles.

■ **camphor** Naturally-occurring organic compound with a penetrating aromatic odour. Alternative names: 2-camphanone, gum camphor.

■ **cane-sugar** Alternative name for **sucrose**.

Cannizzaro reaction In organic chemistry, the formation of an **alcohol** and an acid salt by the reaction between certain **aldehydes** and strong alkalis. It was named after the Italian chemist Stanislao Cannizzaro (1826–1910).

canonical form *See* **mesomerism**.

caprylic acid Alternative name for **octanoic acid**.

carbamide Alternative name for **urea**.

carbanion Transient negatively charged organic ion that has one more electron than the corresponding **free radical**.

carbene Organic radical that contains divalent carbon.

carbide *1.* Chemical compound consisting of carbon and another element; an **acetylide**. *2.* Alternative name for **calcium carbide**.

■ **carbohydrate** Compound of carbon, hydrogen and oxygen that contains a **saccharose** group or its first reaction product, and in which the ratio of hydrogen to oxygen is 2:1 (the same as in water). **Cellulose, starch** and all **sugars** are common carbohydrates. Digestible carbohydrates in the diet are a good source of energy. [3/7/b]

■ **carbolic acid** Alternative name for the antiseptic **phenol**.

■ **carbon** C Non-metallic element in Group IVA of the Periodic Table which exists as several **allotropes** (including **diamond** and **graphite**). It occurs in all living things and its compounds are the basis of **organic chemistry**. It is the principal element in coal and petroleum. Its non-organic compounds include the oxides **carbon monoxide** (CO) and **carbon dioxide** (CO_2), **carbides** and **carbonates**. Carbon is used for making electrodes, brushes for electric motors, carbon fibres and in steel. Diamonds are used as gemstones and industrially as abrasives. Its isotope carbon-14 is the basis of **radiocarbon dating** (*see* **beta decay**). At. no. 6; r.a.m. 12.001. [8/8/c]

■ **carbonate** Salt of **carbonic acid** (H_2CO_3), containing the ion CO_3^{2-}. Carbonates commonly occur as minerals (*e.g.* **calcium carbonate**) and are readily decomposed by acids to produce carbon dioxide. *See also* **hydrogencarbonate**

carbonation Addition of **carbon dioxide** under pressure to a liquid. Carbonation is used in making fizzy drinks, and bottled and canned beer. Carbonated water is known as soda water.

■ **carbon black** Finely divided form of **carbon** obtained by the incomplete combustion or thermal decomposition of natural gas or petroleum oil.

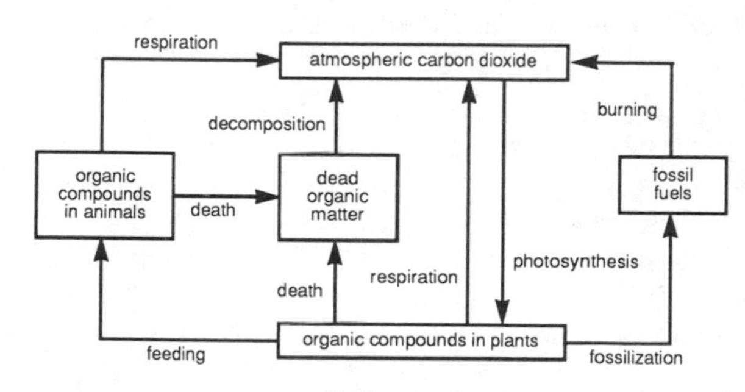

Carbon cycle

■ **carbon cycle** Passage of **carbon** from the air (as **carbon dioxide**) to plants by photosynthesis (forming **sugars** and **starches**), then through the metabolism of animals, to decomposition products which ultimately return to the atmosphere in the form of carbon dioxide. [2/7/a]

■ **carbon dating** *See* **radiocarbon dating**. [8/7/d]

■ **carbon dioxide** CO_2 Colourless gas formed by the combustion of carbon and its organic compounds, by the action of acids on **carbonates**, and as a product of **fermentation** and **respiration**. It is a raw material of photosynthesis. It is used in fire extinguishers, in fizzy (carbonated) drinks, as a coolant in nuclear reactors and, as solid carbon dioxide (dry ice), as a refrigerant. The accumulation of carbon dioxide in the atmosphere creates the **greenhouse effect**. [5/9/a]

carbon disulphide CS_2 Liquid chemical, used as a solvent for oils, fats and rubber and in paint-removers.

■ **carbon fibre** Synthetic carbon produced by charring any spun, felted or woven carbon-containing raw material at temperatures from 700 to 1,800°C. It is used as reinforcement in plastic resins to make high-strength composites.

■ **carbonic acid** H_2CO_3 Acid formed by the combination of **carbon dioxide** and water. Its salts are **carbonates**.

carbonium ion Positively charged fragment that arises from the **heterolytic fission** of a **covalent bond** involving carbon.

■ **carbon monoxide** CO Colourless odourless poisonous gas produced by the incomplete combustion of carbon or its compounds. It is used in the chemical industry as a **reducing agent**. [5/5/a]

carbon tetrachloride Alternative name for **tetrachloromethane**.

carbonyl chloride Alternative name for **phosgene**.

carbonyl compound Chemical containing the radical $=CO$, formed when **carbon monoxide** combines with a metal (*e.g.* nickel carbonyl, $Ni(CO)_4$).

carbonyl group The group $=CO$, as in aldehydes, ketones and carboxyl compounds.

carborundum Form of the hard substance silicon carbide (SiC), used as an abrasive.

carboxyl group The organic group $-COOH$, characteristic of carboxylic acids. Alternative names: carboxy group, oxatyl group.

carboxylic acid Organic acid that contains the carboxyl group, $-COOH$; *e.g.* acetic (ethanoic) acid, CH_3COOH.

carbylamine Alternative name for **isocyanide**.

carnallite Mineral consisting of a hydrated mixture of potassium and magnesium chlorides, mined as a source of potassium and magnesium chloride.

Caro's acid An alternative name for peroxomonosulphuric(VI) acid, H_2SO_5 (persulphuric acid).

■ **casein Protein** that occurs in milk which serves to store **amino acids** as nutrients for the young of mammals. It is the principal constituent of cheese.

■ **cast iron** Hard brittle metal made from remelted **pig iron** mixed with scrap steel, allowed to cool in moulds so that it assumes a definite shape.

■ **catalysis** Action of a **catalyst**.

■ **catalyst** Substance that increases the rate of a chemical reaction without itself undergoing any permanent chemical change. Some reactions take place so slowly that they are virtually impossible without one. *See also* **enzyme**. [7/7a]

■ **catalytic converter** Device in the exhaust system of a vehicle with an internal combustion engine (petrol or diesel), designed to remove gases that contribute to atmospheric pollution. Using platinum or similar catalysts, it converts carbon monoxide to carbon dioxide, nitrogen oxides to nitrogen and unburned hydrocarbon fuel to water and carbon dioxide. [5/5/a]

■ **catalytic cracking** Process of breaking down long-chain **hydrocarbons** into more useful shorter-chain ones, employing a **catalyst**. It is used in an oil refinery to make fuels (*e.g.* petrol, kerosene) from **alkanes** of higher molecular weight. [7/9/a]

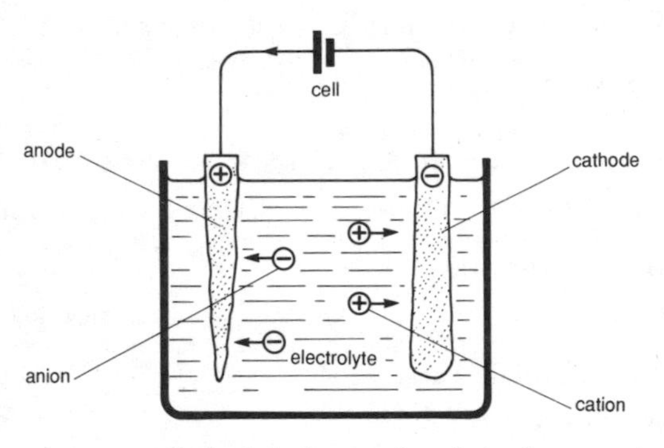

Cathode is the negative electrode

■ **cathode** Negatively-charged electrode in an **electrolytic cell** or **battery**.

■ **cation** Positively charged **ion**, which travels towards the **cathode** during **electrolysis**.

■ **caustic** Describing an **alkaline** substance that is corrosive towards organic matter.

■ **caustic lime** Alternative name for **calcium hydroxide**.

Symbol on containers of caustic chemicals

■ **caustic potash** Alternative name for **potassium hydroxide**.

■ **caustic soda** Alternative name for **sodium hydroxide**.

■ **cell** Alternative name for an **electrolytic cell**.

■ **cellulose** Major **polysaccharide** of plants found in cell walls and in some algae and fungi. It is composed of **glucose** units aligned in long parallel chains, and gives cell walls their strength and rigidity.

cellulose trinitrate Nitric acid **ester** of **cellulose**, used in **plastics**, lacquers and explosives. Alternative names: cellulose nitrate, nitrocellulose.

■ **Celsius scale** Temperature scale on which the **freezing point** of water is 0°C and the **boiling point** is 100°C. It is the same as the formerly used centigrade scale, and a degree Celsius is equal to a unit on the **kelvin** scale. To convert a Celsius temperature to kelvin, add 273.15 (and omit the degree sign). To convert a Celsius temperature to a Fahrenheit one, multiply by 9/5 and add 32. It is named after the Swedish astronomer Anders Celsius (1701–44). [13/4/d]

■ **cement** Any bonding material. The cement used in building to make mortar and concrete consists of powdered roasted rock (*e.g.* Portland stone), which re-forms a solid after absorbing water.

centigrade scale Former name for the **Celsius scale**.

■ **centrifuge** Apparatus for separating substances using sedimentation by spinning a liquid suspension at high speeds, *e.g.* the separation of precipitates in chemical analysis or components of cells in biology. The rate of sedimentation varies according to the particle size of the component.

■ **ceramics** Hard strong materials, produced by firing mixtures containing clay, which have important industrial applications, such as pottery and bricks. Glass and fused **silica** are sometimes also regarded as ceramics. [6/7/a]

cerium Ce Steel-grey metallic element in Group IIIB of the Periodic Table (one of the **lanthanides**). It is used in lighter flints, tracer bullets, catalytic converters for car exhausts and gas mantles. At. no. 58; r.a.m. 140.12.

■ **CFC** Abbreviation of **chlorofluorocarbon**. [5/9/a]

c.g.s. unit Centimetre-gram-second unit, in general scientific use before the adoption of SI units.

■ **chain reaction** Nuclear or chemical reaction in which the products ensure that the reaction continues (*e.g.* nuclear fission, combustion). [8/8/b]

chalk Mineral form of **calcium carbonate**. Blackboard chalk is usually made from **calcium sulphate**.

chamber process Alternative name for **lead-chamber process**.

■ **charcoal** Form of carbon made from incomplete burning of animal or vegetable matter.

■ **Charles's law** At a given pressure, the volume of an ideal gas is directly proportional to its absolute temperature. It was named after the French physicist Jacques Charles (1746–1823). [6/8/a]

chelate Chemical compound in which a central metal ion forms part of one or more organic rings of atoms. The formation of these compounds is useful in many contexts, *e.g.* medicine, in which chelating agents are administered to counteract poisoning by certain heavy metals. They can also act to buffer the concentration of metal ions (*e.g.* iron and calcium) in natural biological systems.

■ **chemical bond** Linkage between atoms or ions within a molecule. A chemical reaction and the input or output of energy are involved in the formation or destruction of a chemical bond. *See also* **covalent bond; ionic bond**. [8/8/c]

■ **chemical combination** Union of two or more chemical substances to form a different substance or substances.

■ **chemical dating** Determination of the age of an archaeological specimen by chemical analysis. *See also* **radiocarbon dating**.

■ **chemical engineering** Manufacture and operation of machines and plant necessary for industrial-scale chemical processing, often to produce other chemicals.

■ **chemical equation** Way of expressing a **chemical reaction** by placing the **formulas** of the **reactants** to the left and those of the **products** to the right, with an equality sign or directional arrows in between. The number of atoms of any particular element are the same on each side of the equation (when the reaction is in equilibrium). *E.g.* the equation for the reaction between two molecules of hydrogen and one molecule of oxygen to form two molecules of water is written as:

$$2H_2 + O_2 \rightarrow 2H_2O.$$ [8/9/a]

■ **chemical equilibrium** Balanced state of a **chemical reaction**, when the concentration of **reactants** and **products** remain constant. *See also* **equilibrium constant**. [7/8/a]

■ **chemical potential** Measure of the tendency of a **chemical reaction** to take place.

■ **chemical reaction** Process in which one or more substances react to form a different substance or substances. [7/4/a]

chemical symbol Letter or pair of letters that stand for an **element** in chemical formulae and equations. *E.g.* the symbols of carbon, chlorine and gold are C, Cl and Au (from the Latin *aurum* = gold). [7/4/a]

■ **chemistry** The study of elements and their compounds, particularly how they behave in **chemical reactions**.

■ **Chile saltpetre** Old name for impure **sodium nitrate**.

chirality Property of a **molecule** that has a carbon atom attached to four different atoms or groups, and which can therefore exist as a pair of optically active **stereoisomers** (whose molecules are mirror images of each other). Commonly called 'handedness', chirality is significant in the biological activity of molecules, in some of which the right-handed version is active and the left-handed version is not, or vice versa (*e.g.* many pheromones).

chloral Alternative name for **trichloroethanal**.

chloral hydrate Alternative name for **trichloroethanediol**.

chlorate Salt of chloric acid ($HClO_3$).

chloride **Binary compound** containing **chlorine**; a salt of hydrochloric acid (HCl).

chlorination *1.* Reaction between **chlorine** and an organic compound to form the corresponding chlorinated compound. *E.g.* the chlorination of benzene produces chlorobenzene, C_6H_5Cl. *2.* Treatment of a substance with **chlorine**; *e.g.* to bleach or disinfect it.

■ **chlorine** Cl Gaseous non-metallic element in Group VIIA of the Periodic Table (the **halogens**), obtained by the electrolysis of sodium chloride (common salt). It is a green-yellow poisonous gas with an irritating smell, used as a disinfectant and bleach and to make chlorine-containing organic chemicals. At. no. 17; r.a.m. 35.453. [6/9/a]

chlorine(I) oxide Alternative name for **dichlorine oxide**.

chlorite Salt of chlorous acid ($HClO_2$).

■ **chlorofluorocarbon** (CFC) **Fluorocarbon** that has chlorine atoms in place of some of the fluorine atoms. Chlorofluorocarbons and fluorocarbons have similar properties, and are used as aerosol propellants and refrigerants. [5/9/a]

chloroform Alternative name for **trichloromethane**.

■ **chlorophyll** Green pigment found in photosynthetic cells (contained in chloroplasts) of green plants. It is the major light-absorbing pigment and the site of the first stage of photosynthesis. [3/6/b]

chloroplatinic acid Solution of platinum chloride used in platinizing glass and ceramics.

chloroprene $CH_2=CCLCH=CH_2$ Colourless liquid organic chemical used (through **polymerization**) to make artificial rubber. Alternative name: 2-chlorobuta-1,3-diene.

■ **cholesterol Sterol** found in animal tissues, a **lipid**-like substance that occurs in blood plasma, cell membranes, nerves and may form gallstones. High levels of cholesterol in the blood are connected to the onset of atherosclerosis, in which fatty materials are deposited in patches on artery walls and can restrict blood flow. Many **steroids** are derived from cholesterol.

■ **chromatography** Method of separating a mixture by carrying it in solution or in a gas stream through an absorbent. The separated substances may be extracted by **elution**. [6/5/c]

chromite Mineral consisting of the oxides of **chromium** and **iron**. Alternative name: chrome iron ore.

■ **chromium** Cr Silver-grey metallic element in Group VIB of the Periodic Table (a **transition element**), obtained mainly from its ore chromite. It is electroplated onto other metals (particularly steel) to provide a corrosion-resistant decorative finish, and alloyed with nickel and iron to make stainless steels; its compounds are used in pigments and dyes. At. no. 24; r.a.m. 51.996.

chromophore Chemical grouping that causes compounds to have colour (*e.g.* the $-N=N-$ group in an **azo dye**).

cinnamic acid $C_6H_5CH=CHCOOH$ White crystalline organic compound with a pleasant odour. Alternative name: 3-phenylpropenoic acid.

■ **citric acid** $C_6H_8O_7$ Hydroxy-tri**carboxylic acid**, present in the juices of fruits. It is important in the energy-generating reactions in cells (Krebs cycle), and is much used as a flavouring.

Claisen condensation Chemical reaction in which two molecules combine to give a compound containing a **ketone** group and an **ester** group.

clathrate Chemical structure in which one atom or molecule is 'encaged' by a structure of other molecules, and not held by **chemical bonds**. *See also* **chelate**.

■ **coal** Black mineral consisting mainly of **carbon**, used as a fuel and source of organic chemicals, and for making **coal gas**, **coal-tar**, **coke** and (by **hydrogenation**) oil. It is the remains of plants from the Carboniferous and Permian periods that have been subjected to high pressures underground. *See also* **peat**. [5/6/c]

■ **coal gas** Fuel gas made by the **destructive distillation** of coal in closed iron retorts, which yields **coke** and **coal-tar** as by-products. The gas consists mainly of hydrogen and methane, with some carbon monoxide.

■ **coal-tar** Thick black oily liquid, consisting mainly of **aromatic compounds**, obtained as a by-product of **coal gas** manufacture. *See also* **tar**.

■ **cobalt** Co Silver-white magnetic metallic element in Group VIII of the Periodic Table, used in alloys to make cutting tools and magnets. The radioactive isotope Co-60 is used in **radiotherapy**. At. no. 27; r.a.m. 58.9332. [3/2/b]

■ **codeine** Pain-killing drug, the methyl derivative of **morphine**.

coenzyme Organic compound essential to catalytic activities of **enzymes** without being utilized in the reaction. Coenzymes usually act as carriers of intermediate products, *e.g.* ATP.

coherent units System of units in which the desired units are obtained by multiplying or dividing base units, with no numerical constant involved. *See also* **SI units**.

■ **coke** Brittle grey-black solid containing about 85 per cent carbon, made by roasting coal in a limited supply of air. It is used as a smokeless fuel and source of carbon (*e.g.* in smelting metal ores).

■ **colloid** Form of matter that consists of small particles, about 10^{-4} to 10^{-6} mm across, dispersed in a medium such as air or water. Common colloids include **aerosols** (*e.g.* fog, mist) and **gels** (*e.g.* gelatin, rubber). A non-colloidal substance is termed a crystalloid.

■ **colorimeter** Instrument for measuring the colour intensity of a medium such as a coloured solution.

complex compound Alternative name for a **co-ordination compound**.

complex ion Cation bonded by means of a **co-ordinate bond**.

■ **compound** Substance that consists of two or more **elements** chemically united in definite proportions by weight. *E.g.* sodium chloride (common salt), NaCl, is a compound of the alkali metal sodium and the halogen gas chlorine. Alternative name: chemical compound. [6/6/b]

■ **concentration** Strength of a mixture or solution. Concentrations can be expressed in very many ways; *e.g.* parts per million (for traces of a substance), percentage (*i.e.* parts per hundred by weight or volume), gm or kg per litre of solvent or per litre of solution, moles per litre of solution (**molarity**), moles per kg of solvent (**molality**), or in terms of normality (*see* **normal**). *See also* **solubility**.

■ **condensation** *1.* Change of a gas or vapour into a liquid or solid by cooling. *2.* Chemical reaction in which two or more small molecules combine to form a larger molecule, often with the elimination of a simple substance such as water. It is used to make **polymers**.

■ **condensation reaction** Chemical reaction in which two or more small molecules combine to form a larger one, often with the elimination of a simpler substance, usually water; *e.g.* acetaldehyde (ethanal), CH_3CHO, condenses with hydroxylamine, H_2NOH, to form an oxime, $CH_3CH{=}NOH$. [7/9/b]

■ **conductor** *1.* Material that allows heat to flow through it by conduction. *2.* Material that allows electricity to flow through it (*e.g.* most metals and their alloys); a conductor has a low resistance. [11/3/a]

■ **configuration** Arrangement in space of the atoms in a molecule.

conformation Particular shape of a molecule that arises through the normal rotation of its atoms or groups about single bonds.

conjugated describing an organic compound that has the single and double triple bonds; *e.g.* buta-1,3-diene, $H_2C{=}CH\text{-}CH{=}CH_2$.

■ **conservation of mass** Principle which states that the products of a purely chemical reaction have the same total mass as the reactants.

■ **constant composition, law of** In any given chemical compound, the same **elements** are always combined in the same proportions by mass.

■ **contact process** Industrial process for the manufacture of **sulphuric acid**, involving the catalytic **oxidation** of **sulphur dioxide** to **sulphur trioxide**. [7/4/b]

control rod Length of **neutron**-absorbing material, *e.g.* boron, cadmium, used to control the rate of fission in a nuclear reactor (by being moved in or out of the core).

co-ordinate bond Type of **covalent bond** that is formed by the donation of a **lone pair of electrons** from one atom to another. Alternative name: dative bond.

co-ordination compound Chemical compound that has **co-ordinate bonds**; *e.g.* potassium ferricyanide, $K_3Fe(CN)_6$. Alternative names: complex, complex compound.

co-ordination number Number of nearest neighbours of an atom or an ion in a chemical compound.

copolymer Polymer built up from two or more different kinds of **monomers**, *e.g.* the hard plastic ABS (acrilonitrile-butadiene-styrene).

■ **copper** Cu Reddish metallic element in Group IB of the Periodic Table (a **transition element**) which occurs as the free metal (native) and in various ores, chief of which is chalcopyrite. The metal is a good conductor of electricity and is used for making wire, pipes and coins. Its chief alloys are **brass** and **bronze**. Its compounds are used as pesticides and pigments. Copper is an important **trace element** in many plants and animals. At. no. 29; r.a.m. 63.546.

copper(I) Alternative name for cuprous in copper compounds.

copper(II) Alternative name for cupric in copper compounds.

copper(II) carbonate $CuCO_3$ Green crystalline compound which occurs (as the **basic salt**) in the minerals azurite and malachite. It is also a component of verdigris, which forms on copper and its alloys exposed to the atmosphere.

copper(II) chloride $CuCl_2$ Brown covalently bonded compound, which forms a green crystalline dihydrate. It is used in fireworks to give a green flame and to remove sulphur in the refining of **petroleum**. Alternative name: cupric chloride.

copper(I) oxide Cu_2O Insoluble red powder, used in rectifiers and as a **pigment**. Alternative names: copper oxide, cuprite, cuprous oxide, red copper oxide.

copper(II) oxide CuO Insoluble black solid, used as a **pigment**. Alternative names: copper oxide, cupric oxide.

copper(II) sulphate $CuSO_4$ White **hygroscopic** compound, which forms a blue crystalline pentahydrate, used as a wood preservative, fungicide, dyestuff and in **electroplating**. Alternative names: blue vitriol, cupric sulphate.

coral Substance containing calcium carbonate that is secreted by various marine organisms (*e.g.* Anthozoa) for support and habitation.

cordite Smokeless propellant explosive prepared from nitrocellulose (**cellulose trinitrate**) and nitroglycerin (**glyceryl trinitrate**).

■ **corrosion** Gradual chemical breakdown, often **oxidation** of metals, by air, water or chemicals. [7/7/d]

corrosive sublimate Alternative name for **mercury(II) chloride**.

■ **corundum** Hard mineral consisting of aluminium oxide (Al_2O_3), whose coloured crystalline forms are ruby and sapphire. It is used as an abrasive (*e.g.* as in emery).

■ **covalent bond** Chemical bond that results from the sharing of a pair of **electrons** between two atoms. *See also* **co-ordinate bond; ionic bond**. [8/8/c]

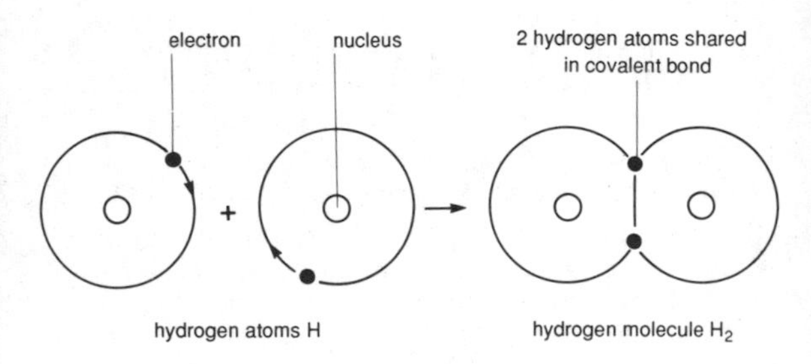

Formation of a covalent bond

■ **covalent compound** Chemical compound in which the atoms are covalently bonded. [8/8/c]

covalent radius Effective radius of an atom involved in a **covalent bond**.

■ **cracking** Industrial process, usually employing heat or a **catalyst**, in which heavy **hydrocarbons** are broken down into lower boiling point fractions, *e.g.* in making petrol from crude oil. It is used in **reforming**. [7/9/a]

cream of tartar Alternative name for **potassium hydrogentartrate**.

■ **creosote** Colourless oily fluid distilled from wood tar; used as a disinfectant. *See also* **creosote oil**.

■ **creosote oil** Dark brown distillation product of **coal-tar**; used as a wood preservative. Alternative names: coal-tar creosote; creosote.

cresol Alternative name for **methylphenol.**

■ **critical mass** Minimum amount of fissile material required to maintain a nuclear **chain reaction.**

■ **cryolite** Na_3AlF_6 Pale grey mineral found mainly in Greenland, used as flux in the electrolytic extraction of **aluminium**. Alternative name: sodium aluminium fluoride.

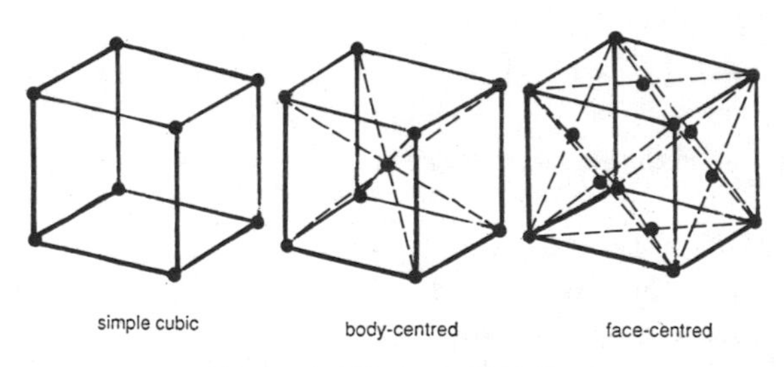

Types of cubic crystal structures

■ **crystal** Solid that has a definite regular shape. Crystals of the same substance have the same specific shapes, reflecting the way its component atoms or ions are regularly arranged in a lattice. *See also* **liquid crystal**.

■ **crystalline** Having the form of crystals, as opposed to being **amorphous**.

■ **crystallization** Process in which crystals form from a molten mass or solution.

■ **crystallography** Study of crystals.

crystalloid Substance that is not a **colloid** and can therefore pass through a semipermeable membrane.

■ **CS gas** $C_6H_4ClCH=C(CN)_2$ White irritant organic compound used as a tear gas for riot control. Alternative name: (2-chlorobenzylidene)-malononitrile.

cupric Bivalent copper. Alternative names: copper(II), copper(2+).

cuprite Mineral form of **copper(I) oxide**.

cuprous Monovalent copper. Alternative names: copper(I), copper(1+).

curium Cm Radioactive element in Group IIIA of the Periodic Table (one of the **actinides**), made by the alpha-particle bombardment of plutonium-239. It has several **isotopes**, with half-lives up to 1.7 × 107 years. At. no. 96; r.a.m. (most stable isotope) 247.

cyanamide $NH=C=NH$. Colourless crystalline compound, used as an industrial source of **ammonia**.

■ **cyanide** Compound containing the cyanide ion (CN^-); a salt of hydrocyanic acid (HCN). All cyanides are highly poisonous.

cyanide, organic *See* **nitrile**.

cyanoferrate See **ferricyanide**; **ferrocyanide**.

cyanogen NCCN Colourless highly poisonous gas made by the action of acids on cyanides.

cyclic Describing the molecule of any (usually organic) compound whose atoms form a ring. **Benzene** and the **cycloalkanes** are simple cyclic compounds. *See also* **acyclic**; **alicyclic**.

cycloalkane Saturated **cyclic** hydrocarbon whose molecule has a ring of carbon atoms, general formula C_nH_{2n}; *e.g.* cyclohexane, C_6H_{12}.

cyclonite $(CH_2NNO_2)_3$ Powerful explosive derived from **hexamine**. Alternative names: hexogen, RDX.

cyclopentadiene C_5H_6 Colourless liquid cyclic hydrocarbon, with a five-membered ring containing two carbon-carbon double bonds. It readily forms the cyclopentadienyl ion $C_5H_5^-$, present in compounds such as **ferrocene**.

cystine $(SCH_2CH(NH_2)COOH)_2$ Dimeric form of the **amino acid** cysteine, found in keratin.

D

dalton Alternative name for **atomic mass unit**.

■ **Dalton's atomic theory** Theory that states that matter consists of tiny particles (atoms), and that all the atoms of a particular **element** are exactly alike, but different from the atoms of other elements in behaviour and mass. The theory also states that chemical action takes place as a result of attraction between atoms, but it fails to account satisfactorily for the volume relationships that exist between combining gases. It was proposed by the British scientist John Dalton (1766–1844). [8/5/a]

Dalton's law of partial pressures In a mixture of gases, the pressure exerted by one of the component gases is the same as if it alone occupied the total volume.

■ **Daniell cell Electrolytic cell** that consists of a zinc **half- cell** and a copper half-cell, usually arranged as a zinc **cathode** and a copper **anode** dipping into an **electrolyte** of dilute sulphuric acid.

dative bond Alternative name for a **co-ordinate bond**.

■ **daughter element** One of the elements produced when an atom divides by **nuclear fission**.

■ **Davy lamp** Alternative name for **safety lamp**.

■ **DDT** Abbreviation of dichlorophenyltrichloroethane. The compound was once much used as an insecticide, but is now banned in most countries because of its high toxicity and the fact that it can pass through food chains without degrading.

deaminase Enzyme that catalyses the removal of an **amino group** ($-NH_2$) from an organic molecule.

deamination Enzymatic removal of an **amino group** ($-NH_2$) from a compound. The process is important in the breakdown of **amino acids** in the liver and kidney. Ammonia formed by deamination is converted to **urea** and excreted.

Debye-Hückel theory Theory that explains variations from the ideal behaviour of **electrolytes** in terms of inter-ionic attraction, and assumes that electrolytes in solution are completely dissociated into charged ions. It was named after the Dutch physicists Peter Debye (1884–1966) and Erich Hückel (1896–).

decane $C_{10}H_{22}$ Colourless liquid hydrocarbon; an **alkane** and tenth member of the methane series.

■ **decant** Carefully pour away the liquid above a precipitate, once it has settled.

■ **decay** *1.* Natural breakdown of an organic substance; **decomposition**. *2.* Breakdown, through **radioactivity**, of a radioactive substance. Such decay is typically exponential. *See also* **half-life**. [8/9/d]

■ **decomposition** *1.* Breaking down of a chemical compound into its component parts, often brought about by heating. *2.* Rotting of a dead organism, often brought about by bacteria or fungi. [7/7/c]

■ **decrepitation** *1.* Crackling sound that is often produced when crystals are heated. It is caused by internal stresses and fracturing. *2.* Fragmentation of powder particles that results from heating.

defect Irregularity in the ordered arrangement of atoms, ions or electrons in a **crystal**.

■ **degradation** Conversion of a molecule into simpler components.

■ **degree of freedom** One of several variable factors, *e.g.* temperature, pressure and concentration, that must be made constant for the condition of a system at equilibrium to be defined.

■ **dehydrating agent** Alternative name for a **desiccant**.

■ **dehydration** Elimination of water from a substance or organism.

dehydrogenase **Enzyme** that catalyses the removal of hydrogen from a compound, and thus causes the compound's **oxidation**.

■ **deionization** Method of purifying or otherwise altering the composition of a solution using **ion exchange**. [5/6/a]

■ **deliquescence** Gradual change undergone by certain substances that absorb water from the atmosphere to become first damp and then aqueous solutions.

delocalization Phenomenon that occurs in certain molecules, *e.g.* **benzene**. Some of the electrons involved in bonding the atoms are not restricted to one particular bond, but are free to move between two or more bonds. The electrical conductivity of metals is due to the presence of delocalized electrons. *See also* **aromaticity**.

delta metal Tough alloy of copper and zinc that contains a small amount of iron.

delta ray Electron that is ejected from an atom when it is struck by a high-energy particle.

dendrite In crystallography, a branching **crystal**, such as occurs in some rocks and minerals.

denitrification Process that occurs in organisms, *e.g.* bacteria in soil, which breaks down **nitrates** and **nitrites**, with the liberation of **nitrogen**.

depolarization Removal or prevention of electrical polarity or **polarization**.

depression of freezing point Reduction of the **freezing point** of a liquid when a solid is dissolved in it. At constant pressure and for dilute solutions of a non-volatile solvent, the depression of the freezing point is directly proportional to the concentration of the solutes.

derivative Compound, usually organic, obtained from another compound.

■ **DERV** Abbreviation of diesel engine road vehicle, another name for **diesel fuel**.

■ **desalination** Removal of salts (mainly sodium chloride) from seawater to produce water pure enough for drinking, irrigation and use in steam turbines and water-cooled machinery. Methods employed include **distillation**, **ion exchange**, **electrodialysis** and **molecular sieves**. [5/6/a]

■ **desiccant** Substance that absorbs water and can therefore be used as a drying agent or to prevent **deliquescence** (*e.g.* anhydrous calcium chloride, silica gel), often used in a **desiccator**. Alternative name: dehydrating agent.

■ **desiccator** Apparatus used for drying substances or preventing **deliquescence**, *e.g.* a closed glass vessel containing a **desiccant**.

desorption *1.* Reverse process to **adsorption**. *2.* Removal of an adsorbate from an **adsorbent** in **chromatography**.

■ **destructive distillation** Heating of a solid (such as coal, wood) in a closed vessel to a high temperature until it decomposes and the products of decomposition (*e.g.* gas, tar) can be carried off as vapour and possibly condensed.

■ **detergent Surfactant** that is used as a cleaning agent. Detergents are particularly useful for cleaning because they lower surface tension and emulsify fats and oils (allowing them to go into solution with water without forming a scum with any of the substances that cause **hardness of water**). **Soaps** act in a similar way, but form an insoluble scum in hard water.

deuterated compound Substance in which ordinary hydrogen has been replaced by **deuterium**.

■ **deuterium** D_2 One of the three isotopes of **hydrogen**, having one neutron and one proton in its nucleus. R.a.m. 2.0141. Alternative name: heavy hydrogen. *See also* **tritium**.

deuterium oxide D_2O Chemical name of **heavy water**.

deuteron Positively charged particle that is composed of one **neutron** and one **proton**; it is the nucleus of a **deuterium** atom. Deuterons are often used to bombard other particles inside a cyclotron.

dextrin Polysaccharide of intermediate chain length produced from the action of **amylases** on **starch**.

dextronic acid Alternative name for **gluconic acid**.

dextrorotatory Describing an **optically active** compound that causes the plane of polarized light to rotate in a clockwise direction. It is indicated by the prefix (+)- or *d*-.

■ **dextrose** Alternative name for **glucose**.

dialysed iron Colloidal solution of iron(III) hydroxide ($Fe(OH)_3$). It is a red liquid, used in medicine.

■ **dialysis** Separation of **colloids** from **crystalloids** using selective diffusion through a semipermeable membrane. It is the process by which globular **proteins** can be separated from low-molecular weight solutes, as in filtering ('purifying') blood in an artificial kidney machine: the membrane retains protein molecules and allows small solute molecules and water to pass through.

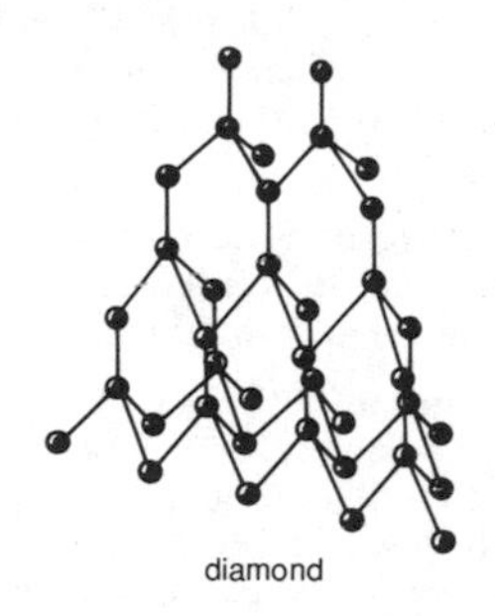

Atomic structure of a diamond

■ **diamond** Colourless crystalline natural **allotrope** of **carbon**, the hardest mineral. Gem-quality diamonds are used in jewellery, and industrial ones for cutting tools and precision instruments such as watches. In practice, most diamonds are not pure, and so have some degree of colour. *See also* **allotropy**. [8/8/c]

■ **diastase** Alternative name for **amylase**.

■ **diatomic** Describing a molecule that is composed of two identical atoms, *e.g.* O_2, H_2 and Cl_2. *See also* **atomicity**.

diatomite Alternative name for **kieselguhr**.

diazo compound Organic compound that contains two adjacent nitrogen atoms, but only one attached to a carbon atom. Formed by **diazotization**, diazo compounds are very important in synthesis, being the starting point of various dyes and drugs.

diazonium compound Organic compound of the type $RN_2^+X^-$, where R is an **aryl group**. The compounds are colourless solids, extremely soluble in water, used for making **azo dyes**. Many of them (particularly the nitrates) are explosive in the solid state.

diazotization Formation of a **diazo compound** by the interaction of sodium nitrite, an inorganic acid, and a primary aromatic amine at low temperatures.

■ **dibasic** Describing an **acid** that contains two replaceable hydrogen atoms in its molecules; *e.g.* carbonic acid, H_2CO_3, sulphuric acid, H_2SO_4. A dibasic acid can form two types of **salts**: a normal salt, in which both hydrogen atoms are replaced by a metal or its equivalent (*e.g.* sodium carbonate, Na_2CO_3), and an **acid salt**, in which only one hydrogen atom is replaced (*e.g.* sodium hydrogensulphate (bisulphate), $NaHSO_4$).

■ **dibromoethane** $C_2H_4Br_2$ Volatile liquid that exists in two isomeric forms. It is added to petrol to remove lead. Alternative name: ethylene dibromide.

dicarboxylic acid Organic acid that contains two **carboxyl groups**.

dichlorine oxide Cl_2O Yellowish-red gas which dissolves in water to produce **hypochlorous acid** (HClO) and explodes to give chlorine and oxygen when heated. Alternative name: chlorine(I) oxide.

dichromate Salt containing the dichromate(VI) ion ($Cr_2O_7^{2-}$), an **oxidizing agent**; *e.g.* potassium dichromate, $K_2Cr_2O_7$. Alternative name: bichromate.

Diels-Alder reaction Addition reaction in organic chemistry in which a 6-membered ring system is formed without elimination of any compounds. It was named after the German chemists Otto Diels (1876–1954) and Kurt Alder (1902–58).

■ **diesel fuel** Type of liquid fuel used in a diesel engine, consisting of **alkanes** (boiling range 200–350°C) obtained from **petroleum**. Alternative name: DERV.

1,1-diethoxyethane Alternative name for **acetal**.

■ **diffusion of gases** Phenomenon by which gases mix together, reducing any concentration gradient to zero; *e.g.* in gas exchange between plant leaves and air.

■ **diffusion of solutions** Free movement of **molecules** or **ions** of a dissolved substance through a solvent, resulting in complete mixing. *See also* **osmosis**.

■ **dihydrate** Chemical (a **hydrate**) whose molecules have two associated molecules of **water of crystallization**; *e.g.* sodium dichromate, $Na_2Cr_2O_7.2H_2O$.

2,3-dihydroxybutanedioic acid Alternative name for **tartaric acid**.

dihydroxypurine Alternative name for **xanthine**.

dihydroxysuccinic acid Alternative name for **tartaric acid**.

diluent Solvent used to reduce the strength of a **solution**.

■ **dilute** *1.* To reduce the strength of a **solution** by adding water or other solvent. *2.* Describing a solution in which the amount of **solute** is small compared to that of the **solvent**.

■ **dilution** Process that involves the lowering of concentration.

dimer Chemical formed from two similar **monomer** molecules.

dimethylbenzene Alternative name for **xylene**.

■ **dinitrogen oxide** N_2O Colourless gas made by heating ammonium nitrate and used as an anaesthetic. Alternative names: nitrogen oxide, nitrous oxide, dental gas, laughing gas.

■ **dinitrogen tetroxide** N_2O_4 Colourless solid which melts at 9°C to give a pale yellow liquid. It is used as an oxidant, *e.g.* in rocket fuel.

diol Organic compound containing two **hydroxyl groups** and having the general formula $C_nH_{2n}(OH)_2$. Diols are thick liquids or crystalline solids, and some have a sweet taste. Ethane-1,2-diol (ethylene glycol) is the simplest diol, widely used as a solvent and as an antifreeze agent. Alternative names: dihydric alcohol, glycol.

dioxan $(CH_2)_2O_2$ Colourless liquid cyclic **ether**. It is inert to many reagents and frequently used in mixtures with water to increase the solubility of organic compounds such as alkyl halides. Alternative name: 1,4-dioxan.

direct dye Dye that does not require a **mordant**.

■ **disaccharide** Sugar with molecules that consist of two **monosaccharide** units linked by **glycoside** bonds; *e.g.* **sucrose, maltose, lactose**.

disintegration constant Probability of a radioactive **decay** of an atomic nucleus per unit time. Alternative names: decay constant, transformation constant.

dislocation Imperfection in a **crystal** lattice.

disodium oxide Alternative name for **sodium monoxide**.

disodium tetraborate Alternative name for **borax**.

■ **displacement reaction** Alternative name for **substitution reaction**.

■ **dissociation** Temporary reversible chemical decomposition of a substance into its component atoms or molecules, which often take the form of **ions**. *E.g.* it occurs when most ionic compounds dissolve in water.

dissociation constant Equilibrium constant of a **dissociation** reaction, and therefore a measure of the affinity of atoms or molecules in a compound.

■ **dissolving** Making a **solution**, usually by adding a solid **solute** to a liquid **solvent**.

■ **distillate** Condensed liquid obtained by **distillation**.

■ **distillation** Method for purification or separation of liquids by heating to the **boiling point**, condensing the vapour, and collecting the **distillate**. Formerly, the method was used to produce distilled water for chemical experiments and processes that required water to be much purer than that in the mains water supply. In this application distillation has been largely superseded by **ion exchange**. [6/5/c]

■ **distilled water** *See* **distillation**.

■ **disulphuric acid** Alternative name for **oleum**.

dithionate Salt derived from **dithionic acid** ($H_2S_2O_6$). Alternative name: hyposulphate.

dithionic acid $H_2S_2O_6$ Strong acid that decomposes slowly in concentrated solutions and when heated. Alternative name: hyposulphuric acid.

dithionite Name that is given to any of the **salts** of **dithionous acid**, all of which are strong reducing agents.

dithionous acid $H_2S_2O_4$ Strong but unstable acid that is found only in solution.

■ **divalent** Capable of combining with two atoms of hydrogen or their equivalent. Alternative name: bivalent.

dl-form Term indicating that a mixture contains the **dextrorotatory** and the **laevorotatory** forms of an optically active compound in equal molecular proportions.

D-lines Pair of characteristic lines in the yellow region of the spectrum of sodium, used as standards in spectroscopy.

■ **DNA** Abbreviation of deoxyribonucleic acid, the long thread-like molecule that consists of a double helix of polynucleotides (combinations of a sugar, organic bases and phosphate) held together by **hydrogen bonds**. DNA is found chiefly in **chromosomes** and is the material that carries the hereditary information of all living organisms (although most, but not all, viruses have only RNA, ribonucleic acid). [4/9/d]

dodecanoic acid Alternative name for **lauric acid**.

dodecylbenzene $C_6H_5(CH_2)_{11}CH_3$ Hydrocarbon of the **benzene** family, important in the manufacture of **detergents**.

■ **dolomite** $CaCO_3.MgCO_3$ Naturally occurring calcium-magnesium carbonate, named after a mountain range in which it is found. Alternative name: pearl spar.

donor Atom that donates both electrons to form a **co-ordinate bond**.

doping Addition of impurity atoms to *e.g.* germanium or silicon to make them into **semiconductors**. Doping with an element of valence 5 (*e.g.* antimony, arsenic, phosphorus) donates electrons to form an *n*-type semiconductor; doping with an element of valence 3 (*e.g.* aluminium, boron, gallium) donates 'holes' to form a *p*-type semiconductor.

dosimeter Instrument used for measuring the dose of **radiation** received by an individual or an area.

■ **double bond Covalent bond** that is formed by sharing two pairs of electrons between two atoms. [8/8/c]

■ **double decomposition** Reaction between two dissolved ionic substances (usually salts) in which the reactants 'change partners' to form a new soluble salt and an insoluble one, which is precipitated. *E.g.* solutions of sodium chloride (NaCl) and silver nitrate ($AgNO_3$) react to form a solution of sodium nitrate ($NaNO_3$) and a precipitate of silver chloride (AgCl).

■ **dry cell** Electrolytic cell containing no free liquid **electrolyte**. A moist paste of ammonium chloride (NH_4Cl) often acts as the electrolyte. Dry cells are used in batteries for torches, portable radios, etc.

■ **dry ice** Solid **carbon dioxide**.

■ **Duralumin** Strong **alloy** of aluminium containing 4% copper and traces of magnesium, manganese and silicon, much used in the aerospace industry.

dynamic isomerism Alternative name for **tautomerism**.

■ **dynamite** High explosive consisting of **nitroglycerine** (and sometimes other explosives) absorbed into the earthy mineral kieselguhr. It was invented by the Swedish chemist Alfred Nobel (1833–96).

dysprosium Dy Silvery metallic element in Group IIIB of the

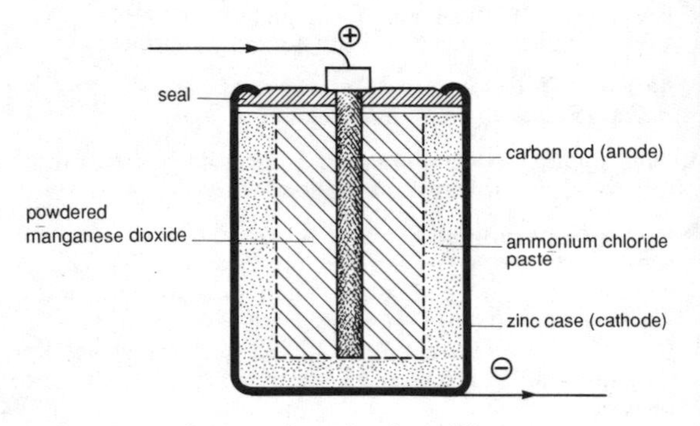

Construction of a dry cell battery

Periodic Table (one of the **lanthanides**), used to make magnets and nuclear reactor **control rods**. At. no. 66; r.a.m. 162.50.

E

Edison accumulator Alternative name for **nickel-iron accumulator**.

EDTA Abbreviation of ethylenediaminetetracetic acid. It is a white crystalline organic compound, used generally as its sodium salt as an analytical reagent and antidote for heavy-metal poisoning, when it forms **chelates**.

■ **effervescence** Evolution of bubbles of a gas from a liquid.

■ **efflorescence** Property of certain crystalline salts that lose **water of crystallization** on exposure to air and become powdery; *e.g.* sodium carbonate decahydrate (washing soda), $Na_2CO_3.10H_2O$. It is the opposite of **deliquescence**.

■ **effusion** Passage of gases under pressure through small holes.

einsteinium Es Radioactive metallic element in Group IIIB of the Periodic Table (one of the **lanthanides**). It has several **isotopes**, with half-lives of up to 2 years. At. no. 99; r.a.m. (most stable isotope) 254.

electrochemical series List of metals arranged in order of their **electrode potentials**. A metal will displace from their salts metals lower down in the series. Alternative name: electromotive series.

■ **electrochemistry** Branch of science that is concerned with the study of electrical chemical energy, such as the effects of electric current on chemicals (particularly **electrolytes**) and the generation of electricity by chemical action (as in an **electrolytic cell**). [7/6/b]

■ **electrode** *1.* Conducting plate (**anode** or **cathode**) that collects or emits electrons from an **electrolyte** during **electrolysis**. *2.* Conducting plate in an **electrolytic cell** (battery), discharge tube or vacuum tube.

electrodeposition Deposition of a substance from an **electrolyte** on to an electrode, as in **electroplating**.

electrode potential Potential developed by a substance in equilibrium with a solution of its ions.

electrodialysis Removal of salts from a solution (often a **colloid**) by placing the solution between two semipermeable membranes, outside which are electrodes in pure solvent.

■ **electrolysis** Conduction of electricity between two **electrodes**, through a solution of a substance (or a substance in its molten state) containing **ions** and accompanied by chemical changes at the electrodes. *See also* **electroplating**. [7/6/b]

■ **electrolysis, Faraday's laws of** *See* **Faraday's laws of electrolysis**.

■ **electrolyte** Substance that in its molten state or in solution can conduct an electric current. [7/6/b]

■ **electrolytic cell** Apparatus that consists of **electrodes** immersed in an **electrolyte**. [7/6/b]

■ **electrolytic dissociation** Partial or complete reversible decomposition of a substance in solution or the molten state into electrically charged **ions**. [7/6/b]

■ **electrolytic refining** Method of obtaining pure metals by making the impure metal the **anode** in an **electrolytic cell** and depositing the pure metal on the **cathode**. [7/7/c]

electrolytic separation Method of separating metals from a solution by varying the applied potential according to the **electrode potentials** of the metals.

electrometallurgy Electrical methods of processing metals. Such methods are used in industry, *e.g.* for **electrodeposition**, **electrolytic refining** and the separation of metals from their ores.

electromotive series Alternative name for **electrochemical series**.

■ **electron** Fundamental negatively-charged subatomic particle (radius 2.81777×10^{-15} m; rest mass 9.109558×10^{-31} kg; charge $1.602\,192 \times 10^{-19+}$ coulombs). Every neutral atom has as many orbiting electrons as there are **protons** in its **nucleus**. A flow of electrons constitutes an electric current. [8/8/a]

electron affinity Energy liberated when an **electron** is acquired by a neutral atom.

■ **electron capture** *1.* Formation of a negative ion through the capture of an **electron** by a substance. *2.* Transformation of a **proton** into a **neutron** in the nucleus of an atom (accompanied by the emission of X-rays) through the capture of an orbital electron, so converting the element into another with an atomic number one less.

■ **electron configuration** *See* **configuration**. [8/8/a]

electron-deficient compound Compound in which there are insufficient electrons to form two-electron **covalent bonds** between all the adjacent atoms *e.g.* boranes.

electron diffraction Method of determining the arrangement of the atoms in a solid, and hence its crystal structure, by the diffraction of a beam of **electrons**.

■ **electron donor** Alternative name for **reducing agent**.

electronegativity Power of an atom in a molecule to attract electrons. For elements arranged in the Periodic Table, it increases up a group and across a period.

electron octet *See* **octet**.

electrophile Electron-deficient ion or molecule that attacks molecules of high electron density. *See also* **nucleophile**.

electrophilic addition Chemical reaction that involves the addition of a molecule to an **unsaturated** organic compound across a double or triple bond.

electrophilic reagent Reagent that attacks molecules of high electron density.

electrophilic substitution Reaction that involves the substitution of an atom or group of atoms in an organic compound. An **electrophile** is the attacking substituent.

electrophoresis Movement of charged **colloid** particles in a solution placed in an electric field.

■ **electroplating** Deposition of thin coating of a metal using **electrolysis**. The object to be plated is the **cathode**, and the plating metal is the **anode**. Metal ions are stripped from the anode, pass through the electrolyte, and are deposited on the cathode. [11/7/b]

electropositive Tending to form positive ions; having a deficiency of electrons.

electrotype Printing plate made by the electrodeposition of copper in a mould (often made from paper or plastic that has been pressed onto metal type).

■ **electrovalent bond** Alternative name for **ionic bond**. [8/8/c]

■ **electrovalent crystal** Crystal in which the **ions** are linked by a bond resulting from electrostatic attraction. Alternative name: ionic crystal.

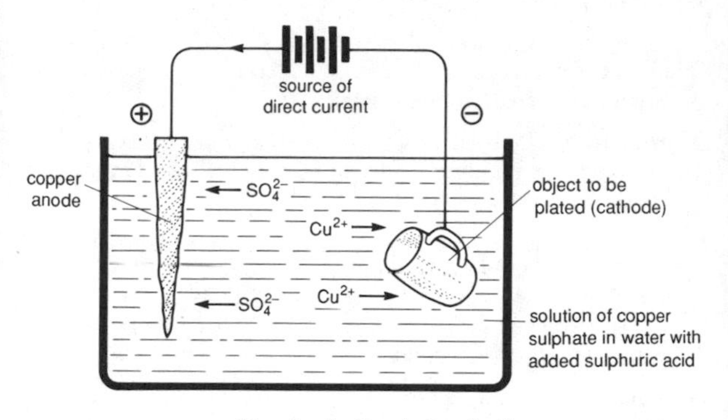

Simple electroplating bath

electrum Naturally occurring **alloy** of gold and up to 40 per cent silver that looks superficially like pure gold.

■ **element** Substance consisting of exactly similar **atoms** of the same atomic number. It cannot be decomposed by chemical action to a simpler substance. Alternative name: chemical element. *See also* **isotope**; **Periodic Table.** [8/7/b]

elementary particle Subatomic particle not known to be made up of simpler particles.

elevation of boiling point Rise in the boiling point of a liquid caused by dissolving a substance in the liquid.

Elinvar Alloy of chromium, iron and nickel, used for making watch hairsprings.

eluate Solution obtained from **elution**.

eluent Solvent used for **elution**, the mobile phase.

elution Removal of an adsorbed substance by washing the adsorbent with a solvent (eluent). The technique is used in some forms of **chromatography**.

■ **emery** Naturally occurring form of **aluminium oxide** (corundum) containing oxides of iron and silica as impurities. It is used as an abrasive.

■ **empirical formula** Chemical formula that shows the simplest ratio between atoms of a molecule. *E.g.* glucose, molecular formula $C_6H_{12}O_6$, and acetic (ethanoic) acid, $C_2H_4O_2$, both have the same empirical formula, CH_2O. *See also* **molecular formula; structural formula**.

■ **emulsion Colloidal** suspension of one liquid dispersed in another.

■ **endothermic** Describing a process in which heat is taken in; *e.g.* in many chemical reactions. *See also* **exothermic**. [7/6/c]

■ **end point** Point at which a chemical reaction is complete, such as the end of a **titration**. *See also* **volumetric analysis**.

■ **energy** *1.* Capacity for doing work, measured in joules. Energy takes various forms: *e.g.* kinetic energy, potential energy, electrical energy, chemical energy, heat, light and sound. All forms of energy can be regarded as being aspects of kinetic or potential energy; *e.g.* heat energy in a substance is the kinetic energy of that substance's molecules. *2.* Fuel or power source (*e.g.* in such expressions as alternative energy, nuclear energy).

■ **energy level** The energy of **electrons** in an atom is not continuously variable, but has a discrete set of values, *i.e.*

energy levels. At any instant the energy of a given electron can correspond to only one of these levels. *See* **Bohr theory**.

enol Organic compound that contains the group C=CH(OH); the alcoholic form of a **ketone**.

■ **enzyme Protein** that acts as a **catalyst** for the chemical reactions that occur in living systems. Without such a catalyst most of the reactions of metabolism would not occur under the conditions that prevail. Most enzymes are specific to a particular **substrate** (and therefore a particular reaction) and act by activating the substrate and binding to it. [7/7/a]

epithermal neutron Neutron that has energy of between 10^{-2} and 10^2 electron volts (eV); a neutron having energy greater than that associated with thermal agitation.

epoxide Organic compound whose molecules include a three-membered oxygen ring (a cyclic ether).

■ **epoxy resin** Synthetic polymeric **resin** with **epoxide** groups. Such resins are used in surface coatings and as adhesives. Alternative name: epoxide resin.

■ **Epsom salt** Alternative name for **magnesium sulphate**.

■ **equation** *See* **chemical equation** and the following article. [8/6/c]

equation of state Any formula that connects the volume, pressure and temperature of a given system, *e.g.* **van der Waals' equation**.

equilibrium constant (K_c) Concentration of the products of a chemical reaction divided by the concentration of the reactants, in accordance with the chemical equation, at a given temperature.

equimolecular mixture Mixture of substances in equal molecular proportions.

equivalence point Theoretical **end point** of a **titration**. *See also* **volumetric analysis**.

equivalent proportions, law of When two **elements** both form chemical compounds with a third element, a compound of the first two contains them in the relative proportions they have in compounds with the third one. *E.g.* Carbon combines with hydrogen to form methane, CH_4, in which the ratio of carbon to hydrogen is 12:4; oxygen also combines with hydrogen to form water, H_2O, in which the ratio of oxygen to hydrogen is 16:2. Carbon and oxygen form the compound carbon monoxide, CO, in which the ratio of carbon to oxygen is 12:16. Alternative name: law of reciprocal proportions.

equivalent weight Number of parts by mass of an element that can combine with or displace one part by mass of hydrogen.

erbium Er Metallic element in Group IIIB of the Periodic Table (one of the **lanthanides**), used in making lasers for medical applications. Its pink oxide is used as a pigment in ceramics. At. no. 68; r.a.m. 167.26.

ergosterol White crystalline **sterol**. It occurs in animal **fat** and in some micro-organisms. In animals it is converted to vitamin D_2 by ultraviolet radiation.

■ **essential amino acid** Any **amino acid** that cannot be manufactured in some vertebrates, including human beings. These acids must therefore be obtained from the diet. They are as follows: arginine, histidine, isoleucine, leucine, lysine, methionine, phenylalanine, threonine, tryptophan and valine.

■ **essential fatty acid** Any **fatty acid** that is required in the diet of mammals because it cannot be synthesized.They include linoleic acid and γ-linolenic acid, obtained from plant sources.

■ **essential oil** Volatile oil with a pleasant odour, obtained from various plants. Such oils are widely used in perfumery.

■ **ester** Compound formed when the hydrogen atom of the **hydroxyl group** in an oxygen-containing acid is replaced by an **alkyl group**, as when a **carboxylic acid** reacts with an **alcohol**; *e.g.* acetic (ethanoic) acid, CH_3COOH, and ethanol (ethyl alcohol), C_2H_5OH, react to form the ester ethyl acetate (ethanoate), $CH_3COOC_2H_5$.

■ **esterification** Formation of an **ester**, generally by reaction between an **acid** and an **alcohol**.

■ **ethanal** Alternative namc for **acetaldehyde**.

ethanal trimer Alternative name for **paraldehyde**.

■ **ethane** C_2H_6 Gaseous **alkane** which occurs with **methane** in natural gas.

ethanedioic acid Alternative name for **oxalic acid**.

ethanediol $HOCH_2CH_2OH$ Syrupy liquid **glycol** (dihydric alcohol), used for making polymers and as an antifreeze. Alternative names: ethylene glycol, glycol.

ethane-propane rubber (EPR) Synthetic rubber prepared by **polymerization** of **ethane** and **propane**.

■ **ethanoate** Alternative name for **acetate**.

■ **ethanoic acid** Alternative name for **acetic acid**.

■ **ethanol** C_2H_5OH Colourless liquid **alcohol**. It is the active constituent of alcoholic drinks; it is also used as a fuel and in the preparation of esters, ethers and other organic compounds. Alternative names: ethyl alcohol, alcohol. [7/5/b]

ethanoyl chloride Alternative name for **acetyl chloride**.

■ **ethene** $CH_2=CH_2$ Colourless gas with a sweetish smell, important in chemical synthesis. Alternative name: ethylene.

ether Organic compound that has the general formula ROR′, where R and R′ are **alkyl groups**. The commonest, diethyl oxide (diethyl ether, or simply ether), $(C_2H_5)O_2$, is a useful volatile solvent formerly used as an anaesthetic.

■ **ethyl acetate** Alternative name for **ethyl ethanoate**.

■ **ethyl alcohol** Alternative name for **ethanol**.

ethylbenzene Alternative name for **styrene**.

ethyl carbamate Alternative name **urethane**.

ethylene Alternative name for **ethene**.

■ **ethylene dibromide** Alternative name for **dibromoethane**.

ethylene glycol Alternative name for **ethanediol**.

■ **ethylene tetrachloride** Alternative name for **tetrachloroethene**.

■ **ethyl ethanoate** Colourless liquid **ester** with a fruity smell, produced by the reaction between ethanol (ethyl alcohol) and ethanoic (acetic) acid, used as a solvent and in medicine. Alternative name: ethyl acetate.

■ **ethyne** Alternative name for **acetylene**.

europium Eu Silvery-white metallic element in Group IIIB of the Periodic Table (one of the **lanthanides**), used in nuclear reactor **control rods**. At. no. 63; r.a.m. 151.96.

eutectic mixture Mixture of substances in such proportions that no other mixture of the same substances has a lower freezing point.

■ **evaporation** Process by which a liquid changes to its vapour. It can occur (slowly) at a temperature below the boiling point, but is faster if the liquid is heated and fastest when the liquid is boiling. [8/7/a]

excess electron Electron that is added to a **semiconductor** from a **donor** impurity. *See also* **doping**.

excitation Addition of energy to a system, such as an atom or nucleus, causing it to transfer from its **ground state** to one of higher energy.

excitation energy Energy required for **excitation**.

excited state Energy state of an atom or molecule that is higher than the **ground state**, resulting from **excitation**. *See also* **energy level**.

exclusion principle Alternative name for the **Pauli exclusion principle**.

■ **exothermic** Describing a process in which heat is evolved. [7/6/c]

■ **expansion of gas** Increase in volume of an ideal gas is at the rate of $^1/_{273}$ of its volume at 0°C for each degree rise in temperature. *See* **Charles' law**.

■ **explosion** Rapid production of heat and pressure from a chemical or nuclear reaction, accompanied by the evolution of large volumes of gas and a destructive shock wave.

extrinsic semiconductor **Semiconductor** that has its conductivity increased by the introduction of tiny, but controlled, amounts of certain impurities. *See* **doping**.

F

face-centred cube Crystal structure that is cubic with an **atom** or **ion** at the centre of each of the six faces of the cube in addition to the eight at its corners.

■ **Fahrenheit scale** Temperature scale on which the freezing point of water is 32°F and the boiling point 212°F. It was named after the German physicist Gabriel Fahrenheit (1686–1736). *See also* **Celsius scale**. [13/4/d]

Fajans' rules Set of rules that state when **ionic bonds** are likely in a chemical compound (as opposed to covalent bonds). They were named after the Polish chemist Kasimir Fajans (1887–1975).

■ **fall-out** Radioactive substances that fall to Earth from the atmosphere after a nuclear explosion.

■ **faraday** (F) Unit of electric charge, equal to the quantity of electric charge that during **electrolysis** liberates one gram equivalent of an element. It has the value 9.65×10^4 coulombs per gram-equivalent.

■ **Faraday's laws of electrolysis** *1.* The quantitative amount of chemical decomposition that takes place during **electrolysis** is proportional to the electric current passed. *2.* The amounts of substances liberated during electrolysis are proportional to their chemical equivalent weights. [11/7/b]

fast neutron Neutron produced by **nuclear fission** that has lost little of its energy and travels too fast to produce further fission and sustain a **chain reaction** (unlike a slow, or thermal, neutron).

fat *See* **fats and oils**.

■ **fats and oils** Naturally occurring **esters** (of **glycerol** and **fatty acids**) that are used as energy-storage compounds by animals and plants. They are hydrocarbons and members of a larger class of naturally occurring compounds called **lipids**.

■ **fatty acid** Monobasic **carboxylic acid**, an essential constituent of **fats and oils**. The simplest fatty acids are formic (methanoic) acid, HCOOH, and acetic (ethanoic) acid, CH_3COOH.

■ **Fehling's solution** Test reagent consisting of two parts: a solution of copper(II) sulphate and a solution of potassium sodium tartrate and sodium hydroxide. When the two solutions are mixed, an alkaline solution of a soluble copper(II) complex is formed. In the presence of an **aldehyde** or reducing **sugar**, a pink-red precipitate of copper(I) oxide forms. It was named after the German chemist Hermann Fehling (1812–85). *See also* **Fehling's test**.

■ **Fehling's test** Test for an **aldehyde** group or **reducing sugar**, indicated by the formation of copper(I) oxide as a pink-red precipitate with **Fehling's solution**.

■ **feldspar** Any of a large group of igneous crystalline rocks, chiefly silicates of aluminium with potassium, sodium and calcium. Feldspar is used in making porcelain, tiles and glazes. Alternative name: felspar.

■ **felspar** Alternative name for **feldspar**.

■ **fermentation** Energy-producing breakdown of organic compounds by micro-organisms (in the absence of oxygen); *e.g.* the breakdown of **sugar** by yeasts into ethanol, carbon dioxide and organic acids. Fermentation is a type of **anaerobic** respiration. [7/5/b]

fermium Fm Radioactive element in Group IIIB of the Periodic Table (one of the **actinides**). It has several **isotopes**, with half-lives of up to 95 days. At. no. 100; r.a.m. (most stable isotope) 257.

ferric Trivalent iron. Alternative names: iron(III), iron(3+).

ferricyanide $[Fe(CN)_6]^{3-}$ Very stable complex ion of iron(III). A solution of the potassium salt gives a deep blue precipitate (Prussian blue) in the presence of iron(II) (ferrous) ions. Alternative name: hexacyanoferrate(III).

ferrite Non-conducting ceramic material that exhibits ferrimagnetism; general formula MFe_2O_4, where M is a divalent metal of the **transition elements**. Ferrites are used to make powerful magnets in radars and other high-frequency electronic apparatus, such as computer memories.

ferrocene $C_{10}H_{10}Fe$ Orange organometallic compound, whose molecules consist of an iron atom 'sandwiched' between two molecules of **cyclopentadiene**. Alternative name: dicyclopentadienyliron.

ferrocyanide $[Fe(CN)_6]^{4-}$ Very stable complex ion of iron(II). Alternative name: hexacyanoferrate(II).

ferrous Bivalent iron. Alternative names: iron(II), iron(2+).

■ **fertilizer** Substance used to increase the fertility of soil. Natural, or organic, fertilizers consist of animal or plant residues and are usually called manures. Artificial, or inorganic, fertilizers supply nitrogen (in the form of compounds such as ammonium nitrate, ammonium sulphate and ammonium phosphate), and sometimes also phosphorus and potassium (potash). They are specified in terms of their NPK (nitrogen, phosphorus, potassium) content. [2/5/c]

Feulgen's test Test for the presence of **DNA**

(deoxyribonucleic acid); *e.g.* in tissue sections by staining, which gives a purple colour. It was named after the German chemist R. Feulgen (1884–1955).

■ **filler** Inert substance added to paper, paints, plastics and rubbers during manufacture to increase their bulk or weight, or otherwise modify their properties.

■ **filter pump** Simple vacuum pump in which a jet of water draws air molecules from the system. It can produce only low pressures and is commonly used to increase the speed of **filtration** by drawing through the **filtrate**.

■ **filtrate** Liquid obtained after **filtration**.

■ **filtration** Method of separating a suspended solid from a liquid by passing the mixture through a porous medium, *e.g.* filter paper or glass wool, through which only the liquid passes.

■ **fine chemical** Chemical produced in pure form and in small quantities.

■ **fireclay** Clay that contains large amounts of **alumina** and **silica**, and is capable of withstanding high temperatures. It is used in the production of refractory brick, furnace linings, etc.

■ **firedamp** Explosive mixture of air and **methane** found in coal mines.

■ **fission** Splitting. In atomic physics, disintegration of an atom into parts with similar masses, usually with the release of energy and one or more neutrons (*see* **nuclear fission**). [8/8/b]

■ **fission product Isotope** produced by **nuclear fission**, with a mass equal to roughly half that of the fissile material. [8/8/b]

■ **fixation of nitrogen** Part of the **nitrogen cycle** that involves the conversion and eventual incorporation of atmospheric nitrogen into compounds that contain nitrogen. Nitrogen fixation in nature is carried out by nitrifying soil bacteria or blue-green algae (Cyanophyta) in the sea. Soil bacteria may exist symbiotically in the root nodules of leguminous plants or they may be free-living. Small amounts of nitrogen are also fixed, as **nitrogen monoxide** (nitric oxide), by the action of lightning. [2/7/a]

■ **fixing, photographic** Process for removing unexposed silver halides after development of a photographic emulsion. It involves dissolving the silver salts by immersing the developed film in a fixing bath consisting of a solution of sodium or ammonium thiosulphate.

■ **flame test** Qualitative chemical test in which an element in a substance is identified by the characteristic colour it imparts to a Bunsen burner flame.

■ **flash point** Lowest temperature at which a substance or a mixture gives off sufficient vapour to produce a flash on the application of a flame.

■ **flint** Natural crystalline form of **silica** that is used as an abrasive. It was important in the Stone Age as a material from which to make tools.

■ **flocculation** Coagulation of a finely divided precipitate into larger particles.

■ **flotation process** Method of concentrating ores by making the ore float on detergent-produced froth in a tank of liquid. Particles adhering to the bubbles are removed with the froth, and thus separated from the materials remaining in the slurry. Alternative name: froth flotation.

■ **fluid** Form of matter that can flow; thus both gases and liquids are fluids. Fluids can offer no permanent resistance to changes of shape. Resistance to flow is manifest as viscosity.

fluorescein $C_{20}H_{12}O_5$ Orange-red powder which dissolves in alkalis to give a green fluorescent solution. It is used as a chemical marker and for dyeing textiles. Alternative name: resorcinolphthalein.

■ **fluoridation** Addition of inorganic fluorides to drinking water to combat dental decay.

■ **fluoride** Compound containing **fluorine**; salt of hydrofluoric acid (HF).

fluorination Replacement of atoms, usually hydrogen, in an organic compound by fluorine.

■ **fluorine** F Gaseous nonmetallic element in Group VIIA of the Periodic Table (the **halogens**). A pale green-yellow poisonous gas, it is highly reactive and the most electronegative element, occurring in fluoride minerals such as **fluorspar**. It is used, as the gaseous uranium(VI) fluoride (UF_6), in the separation by diffusion of uranium **isotopes** and in making **fluorocarbons**. Inorganic **fluorides** are added to water supplies to combat tooth decay. At. no. 9; r.a.m. 18.9984. [6/9/a]

■ **fluorite** Alternative name for **fluorspar**.

■ **fluorocarbon** Very stable organic compound in which some or all of the hydrogen atoms have been replaced by fluorine. Fluorocarbons are used as solvents, aerosol propellants and refrigerants. Their use is being limited because they have been implicated in damage to the ozone layer of the atmosphere. *See also* **Freons**.

■ **fluorspar** CaF_2 Naturally occurring calcium fluoride, used as a flux in glass and as a component of certain cements. Alternative name: fluorite.

■ **flux** *1.* Substance added to help fusion. *2.* Substance used in metallurgy to combine with unwanted materials and cause them to flow so that they can be removed from the metal as a slag.

folic acid Water-soluble B-group **vitamin**. Its deficiency leads to anaemia and it is important in the formation of various **coenzymes**, which are in turn essential for growth and reproduction of cells.

■ **formaldehyde** HCHO Colourless pungent organic gas, an **aldehyde**, which is readily soluble in water. It is used as a disinfectant and in the manufacture of plastics. Alternative name: methanal.

■ **formalin** 40% solution of **formaldehyde** in water.

■ **formic acid** HCOOH Simplest **carboxylic acid**, made commercially by the catalytic combination of carbon monoxide and superheated steam. It occurs naturally in ant venom and nettles. Alternative name: methanoic acid.

■ **formula** Chemical composition of a substance indicated by the symbols of each element present in it and subscripts that show the number of each type of atom involved; *e.g.* the formula for water is H_2O, that for potassium dichromate is $K_2Cr_2O_7.2H_2O$. *See also* **empirical formula; molecular formula**. [8/9/a]

■ **fossil fuel** Mineral fuel that is made from the remains of living organisms, *e.g.* **coal, natural gas, petroleum**. [7/5/a]

fractional crystallization Separation of mixtures of substances by the repeated crystallization of a solution, each time at a lower temperature.

■ **fractional distillation** Separation of a number of liquids with different boiling points by **distillation** and collecting separately the liquids that come off at different temperatures. Alternative name: fractionation.

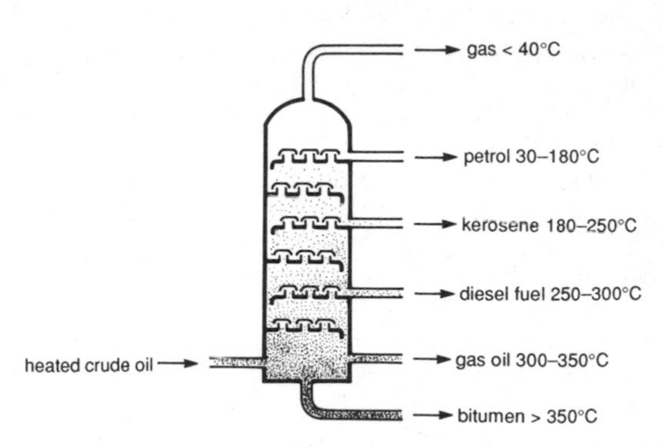

Fractionating column for crude oil

fractionating column Long vertical tube containing bubble-caps, sieve plates, or various irregular packing materials, used for industrial **fractional distillation**.

fractionation Alternative name for **fractional distillation**.

francium Fr Radioactive metallic element in Group IA of the Periodic Table (the **alkali metals**), made by proton bombardment of thorium. It has several **isotopes** with half-lives of up to 22 min. At. no. 87; r.a.m. (most stable isotope) 223.

Frasch process Process for the extraction of sulphur from underground deposits. It involves sinking three concentric tubes to the deposits. Superheated water is pumped down the outer tube to melt the sulphur, and when hot compressed air is injected down the central tube molten sulphur is forced up the remaining tube. It was named after the German chemist Herman Frasch (1851–1914). [7/4/b]

free electron Electron free to move from one atom or molecule to another under the influence of an electric field. Movement of free electrons enables a **conductor** to carry an electric current.

free radical Intermediate and highly reactive molecule that has an unpaired electron and so easily forms a chemical bond.

freezing mixture Mixture of two substances that absorbs heat and can be used to produce a temperature below 0°C. *See also* **eutectic mixture**.

■ **freezing point** Temperature at which a liquid solidifies. Alternative name: solidification point. *See also* **melting point**. [13/4/d]

■ **French chalk** Powdered **talc**, used as a filler and lubricant.

Frenkel defect Crystal disorder that occurs when an ion occupies a vacant interstitial site, leaving its proper site empty.

■ **Freon** Trade name for certain **fluorocarbons** and **chlorofluorocarbons** derived from methane and ethane. They are used as refrigerants.

Friedel-Crafts reaction Chemical reaction that introduces an **alkyl** or **acyl group** into a **benzene** ring, in the presence of a catalyst such as aluminium(III) chloride, boron trifluoride or hydrogen fluoride. It was named after two chemists, the Frenchman Charles Friedel (1832–99) and the American James Crafts (1839–1917).

■ **froth** Collection of fairly stable small bubbles in a liquid produced by shaking or aeration. Alternative name: foam.

froth flotation Process for the separation of finely divided materials in which a slurry is caused to froth by the addition of a foaming agent. Particles adhering to the bubbles are removed with the froth, and thus separated from the materials remaining in the slurry.

fructose $C_6H_{12}O_6$ Fruit sugar, a **monosaccharide** carbohydrate (**hexose**) found in sweet fruits and honey. Alternative name: laevulose. [3/6/b]

fuel cell Device that uses the oxidation of a liquid or gaseous fuel to produce electricity. Thus the chemical energy of the fuel is converted directly to electrical energy, without the intermediate stages of a heat engine and a generator. *E.g.* hydrogen and oxygen flowing over porous platinum electrodes immersed in hot potassium hydroxide solution react to produce water and generate a voltage across the electrodes.

fuller's earth Green, blue or yellow-brown clay-like mineral, used for decolorizing solutions and oils. Alternative name: montmorillonite.

fuming sulphuric acid Alternative name for **oleum**.

functional group Atom or group of atoms that cause a chemical compound to behave in a particular way; *e.g.* the functional group in alcohols is the $-OH$ (hydroxyl) group.

furan C_4H_4O Oxygen-containing **heterocyclic** liquid organic compound. Alternative name: furfuran.

furanose Any of a group of **monosaccharide** sugars (**pentose sugars**) whose molecules have a five-membered heterocyclic ring of four carbon atoms and one oxygen atom. *See also* **pyranose**.

furfuran Alternative name for **furan**.

fusion reaction *See* **nuclear fusion**; **thermonuclear reaction**.

G

gadolinium Gd Silvery-white metallic element in Group IIIB of the Periodic Table (one of the **lanthanides**), which becomes strongly magnetic at low temperatures. At. no. 64; r.a.m. 157.25.

galactose $C_6H_{12}O_6$ **Monosaccharide** sugar that occurs in milk and in certain gums and seaweeds as the **polysaccharide** galactan.

■ **galena** Widely distributed metallic-grey mineral that consists mainly of lead sulphide (PbS); it is the principal ore of **lead**. [9/3/d]

gallium Ga Blue-grey metallic element in Group IIIa of the Periodic Table, used in low-melting-point alloys and thermometers. Gallium arsenide is an important **semiconductor**, used also in lasers. At. no. 31; r.a.m. 69.72.

■ **galvanic cell** Alternative name for a **voltaic cell**.

■ **galvanized iron** Iron or, usually, steel coated with zinc (by dipping or electroplating) to prevent it going rusty. [7/7/d]

galvanizing Method of protecting a metal (*e.g.* iron or steel) from corrosion by covering it with a thin layer of **zinc** through dipping or electroplating.

gamma iron **Iron** with a **face-centred cubic** structure. It is non-magnetic.

■ **gamma radiation** Penetrating form of electromagnetic radiation of shorter wavelength than **X-rays**, produced, *e.g.* during the decay of certain **radio-isotopes**. [8/9/b]

gamma ray High-energy photons that make up **gamma radiation**.

■ **gas** Form (phase) of matter in which the atoms and molecules move randomly with high speeds, occupy all the space available, and are comparatively far apart; a vapour. A liquid heated above its boiling point changes into a gas.

gas chromatography Method of analysing mixtures of substances. The sample is volatilized and then introduced into a column containing the stationary phase (a solid or a non-volatile liquid on an inert support), and an inert carrier gas (*e.g.* argon) is passed through the column. Components of the mixture are removed from the column by the carrier gas at different rates. A detector measures the conductivity of the gas leaving the column, which is recorded on a chart as a series of peaks corresponding to each of the components. The chart is calibrated by passing samples of known composition through the machine.

gas-cooled reactor Nuclear reactor in which the cooling medium is a gas, usually carbon dioxide.

gas equation For n moles of a gas, $pV = n\mathrm{R}T$, where p = pressure, V = volume, n = number of moles, R = the gas constant and T = absolute temperature.

■ **gas laws** Relationships between pressure, volume and temperature of a gas. The combination of Boyle's, Charles' and Gay-Lussac's laws is the **gas equation**. [6/8/a]

gas-liquid chromatography (GLC) Type of **gas chromatography** in which the column contains a non-volatile liquid on an inert support.

■ **gas oil** One of the main fractions of crude oil, used as fuel oil. Alternative name: heavy oil.

■ **Gay-Lussac's law of volume** When gases react their volumes are in a simple ratio to each other and to the volume of products, at the same temperature and pressure. It was named after the French chemist and physicist Joseph Gay-Lussac (1778–1850). [6/8/a]

■ **gel** Jelly-like colloidal solution. *See* **colloid**.

■ **gelatin** Protein extracted from animal hides, skins and bones, which forms a stiff jelly when dissolved in water. It is used as a clarifying agent, in foodstuffs, and in the manufacture of adhesives. Alternative name: gelatine.

gel filtration Type of **chromatography** in which compounds are separated according to their molecular size. Molecules of the components of a mixture penetrate the surface of an inert porous material in proportion to their size. Alternative name: gel permeation.

■ **gelignite** Explosive made from a mixture of nitroglycerine (glyceryl trinitrate), gun-cotton (nitrocellulose, cellulose trinitrate), sodium nitrate and wood pulp. Alternative name: blasting glycerin.

■ **general formula** Expression representing the common chemical **formula** of a group of compounds. *E.g.* C_nH_{2n+2} is the general formula for an **alkane**. A series of compounds of the same general formula constitute a **homologous series**.

geometric isomerism Form of **stereoisomerism** that results from there being no free rotation about a bond between two **atoms**. Groups attached to each atom may be on the same side of the bond (the *cis*-isomer) or on opposite sides (the *trans*-isomer). Alternative name: cis-trans isomerism.

germanium Ge Grey-white semimetallic element in Group IVA of the Periodic Table, which occurs in some silver ores.

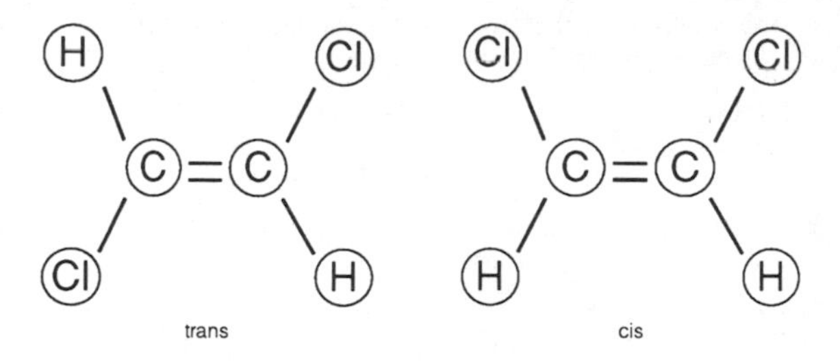

Geometric isomerism

It is an important **semiconductor**, used for making solid-state diodes and transistors. Its oxide is used in optical instruments and infra-red cameras. At. no. 32; r.a.m. 72.59.

German silver Alloy of copper, nickel and zinc, used in coinage and as a base metal for silver-plated articles. Alternative name: nickel silver.

Gibbs free energy (G) Measure of the energy that would be liberated or absorbed during a **reversible process**. $G = H - TS$, where H is heat content, T thermodynamic temperature and S entropy. Alternative name: Gibbs function.

glacial Describing a compound of ice-like crystalline form, especially that of the solid form of a liquid; *e.g.* **glacial acetic** (ethanoic) **acid**.

glacial acetic acid Crystalline form of **acetic** (ethanoic) **acid** below its freezing point.

■ **glass** Hard brittle amorphous mixture of the silicates of sodium and calcium, or of potassium and calcium. In some particularly strong or heat-resistant forms of glass, boron replaces some of the atoms of silicon. Glass is usually transparent or translucent. [6/7/a]

glass ceramic Material that consists of lithium and magnesium aluminium silicates. It is thermally stable, and used for making ovenware. *See also* **ceramics**.

■ **glass electrode** Glass membrane electrode used to measure hydrogen ion concentration or **pH**.

■ **glass wool** Material that consists of fine glass fibres, used in filters, as a thermal insulator and for making fibre-glass.

■ **Glauber's salt** Alternative name for **sodium sulphate**, named after the German physicist Johann Glauber (1603–68).

gluconic acid $CH_2OH(CHOH)_4COOH$ Soluble crystalline organic acid, an oxidation product of **glucose**, used in paint strippers. Alternative name: dextronic acid.

■ **glucose** $C_6H_{12}O_6$ **Monosaccharide** carbohydrate, a soluble colourless crystalline **sugar** (hexose) which occurs abundantly in plants. It is the principal product of photosynthesis and source of energy in animals (it is a product of carbohydrate digestion and is the sugar in blood), used in the manufacture of confectionery and the production of beer. Its natural polymers include **cellulose** and **starch**. Alternative names: dextrose, grape sugar. [3/6/b]

glucoside *See* **glycoside**.

■ **glue** *1.* Adhesive obtained by extracting bones. *2.* Any adhesive made by dissolving a substance such as rubber or plastic in a volatile **solvent**.

gluon Subatomic particle of the type believed to hold **quarks** together.

glyceride Ester of **glycerol** with an organic acid. The most important glycerides are **fats and oils**.

glycerol $HOCH_2CH(OH)CH_2OH$ Colourless sweet syrupy liquid, a trihydric **alcohol** that occurs as a constituent of **fats and oils** (from which it is obtained). It is used in foodstuffs, medicines and in the preparation of alkyd **resins** and nitroglycerine (glyceryl trinitrate). Alternative names: glycerin, glycerine, propan-1,2,3-triol.

glyceryl trinitrate Alternative name for **nitroglycerine**.

glycine $CH_2(NH_2)CO_2H$ Simplest **amino acid**, found in many **proteins** and certain animal excretions. It is a precursor in the biological synthesis of **purines, porphyrins** and **creatine**. It is also a component of **glutathione** and the bile salt **glycocholate**. It acts as a **neurotransmitter** at inhibitory nerve **synapses** in vertebrates. Alternative names: aminoacetic acid, aminoethanoic acid.

glycogen Polysaccharide carbohydrate, the main energy store in the liver and muscles of vertebrates ('animal starch'), also found in some algae and fungi. **Amylase** enzymes convert it to **glucose**, for use in metabolism.

■ **glycol** Alternative name for any **diol** or, specifically, ethylene glycol (**ethanediol**).

glycolipid Member of a family of compounds that contain a **sugar** linked to **fatty acids**. Glycolipids are present in higher plants and neural tissue in animals. Alternative names: glycosylacylglycerols, glycosyldiacylglycerols.

glycoprotein Member of a group of compounds that contain **proteins** attached to **carbohydrate** groups. They include blood glycoproteins, some **hormones** and **enzymes**.

glycoside Compound formed from a **monosaccharide** in which an alcoholic or phenolic group replaces the first hydroxyl group. If the monosaccharide is **glucose**, it is termed a glucoside.

■ **gold** Au Yellow metallic element in Group IB of the Periodic Table. It occurs as the free metal (native) in lodes and placer (alluvial) deposits. Most gold is held in currency reserve stocks, although some is used (usually as an alloy) in coinage, jewellery, dentistry and in electroplating electronic circuits and components. At. no. 79; r.a.m. 196.967.

Graham's law Velocity of diffusion of a gas is inversely proportional to the square root of its density. It was named after the British chemist Thomas Graham (1805–69).

■ **gram** Unit of mass in the metric system. Alternative name: gramme.

■ **gram molecule** Molecular weight of a substance in grams; 1 mole.

grape sugar Alternative name for **glucose**.

■ **graphite** Soft black natural **allotrope** of **carbon**. It is used as a **moderator** in nuclear reactors, as a lubricant, in paints, in pencil 'leads' and as a coating for foundry moulds. [8/8/c]

gravimetric analysis Quantitative chemical analysis made ultimately by weighing substances.

■ **greenhouse effect** Overheating of the Earth's atmosphere resulting from atmospheric pollution, particularly the build-up of carbon dioxide (CO_2), which absorbs and thus traps

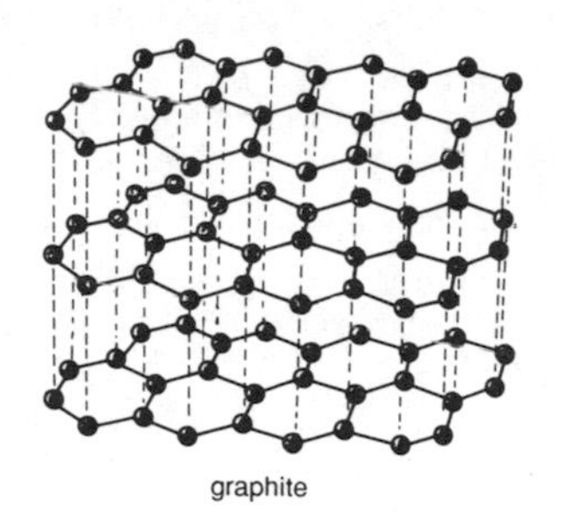

Atomic structure of graphite

some of the solar radiation reflected from the Earth's surface. [3/9/a]

Grignard reagent Member of a class of **organometallic compounds** that have the general formula RMgX, where R = an **alkyl group** and X = Br or I. The reagents are very reactive, giving rise to the highly nucleophilic radical R^-. They are important in organic synthesis. They were named after the French chemist Victor Grignard (1871–1935).

ground state Lowest energy state of an atom or molecule, from which it can be raised to a higher energy state by **excitation**. *See also* **energy level**.

■ **group** Column, or vertical row, of elements in the Periodic Table (horizontal rows are periods). The group number (I to

VIII and 0) indicates the number of **electrons** in the atom's outermost shell. [6/7/c]

guanidine $HN=C(NH_2)_2$ Strongly basic organic compound, used in the manufacture of explosives. Alternative name: imido-urea.

guanine $C_5H_5N_5O$ Colourless crystalline organic base (a **purine** derivative) that occurs in **DNA**.

gum camphor Alternative name for **camphor**.

■ **gypsum** $CaSO_4.2H_2O$ Very soft calcium mineral, used in making cement and plasters. Alternative name: calcium sulphate dihydrate.

H

■ **Haber process** Industrial process for making **ammonia** from hydrogen and atmospheric nitrogen at high temperature and pressure in the presence of a **catalyst**. The hydrogen is obtained by the **Bosch process** or from **synthetic gas**. It was named after the German chemist Fritz Haber (1868–1934). Alternative name: Haber-Bosch process. [7/8/a]

hadron Any **elementary particle** that interacts strongly with other particles, including **baryons** and **mesons**.

■ **haematite** Mineral containing iron(III) oxide, one of the principal ores of iron. One form, kidney ore, occurs as distinctive brown crystals. [9/2/d] [9/3/d]

hafnium Hf Silvery metallic element in Group IVB of the Periodic Table (a **transition element**), used to make **control rods** for nuclear reactors. At. no. 72; r.a.m. 178.49.

hahnium Ha Element no. 105 (a **post-actinide**). It is a radioactive metal with short-lived **isotopes**, made in very small quantities by bombardment of an **actinide** with atoms of an element such as carbon or oxygen.

half-cell Half of an **electrolytic cell**, consisting of an electrode immersed in an **electrolyte**.

■ **half-life** *1.* Time taken for something whose decay is exponential to reduce to half its value. *2.* More specifically, time taken for half the nuclei of a **radioactive** substance to decay spontaneously. The half-life of some unstable substances is only a few seconds or less, whereas for other substances it may be thousands of years; *e.g.* lawrencium has a half-life of 8 seconds, and the **isotope** plutonium-239 has a half-life of 24,400 years. *See also* **radiocarbon dating**. [8/9/d]

■ **halide** Binary compound containing a **halogen**: a fluoride, chloride, bromide or iodide.

■ **halite** Naturally occurring form of **sodium chloride**. Alternative names: common salt, rock salt.

haloalkane Alternative name for **halogenoalkane**.

haloform Organic compound of the type CHX_3, where X is a **halogen** (chlorine, bromine or iodine); *e.g.* chloroform (trichloromethane), iodoform (tri-iodomethane). The compounds are prepared by the action of the halogen on heating with **ethanol** in the presence of sodium hydroxide.

■ **halogen** Element in Group VIIA of the Perodic Table: **fluorine**, **chlorine**, **bromine**, **iodine** or **astatine**. [6/9/a]

■ **halogenation** Chemical reaction that involves the addition of a **halogen** to a substance; *e.g.* the chlorination of benzene (C_6H_6) to form chlorobenzene (C_6H_5Cl).

halogenoalkane **Halogen** derivative of an **alkane**, general formula $C_nH_{2n+1}X$, where X is a **halogen** (fluorine, chlorine, bromine or iodine); *e.g.* chloromethane (methyl chloride), CH_3Cl. Alternative names: haloalkane, alkyl halide; monohalogenoalkane.

■ **hardness of water** Property of water that prevents it forming a lather with soap because of the presence of dissolved compounds of calcium or magnesium. Such compounds form an insoluble scum with soap, and fur up kettles and hot-water pipes. Water with little or no hardness is termed soft. [5/6/a]

heat of activation Difference between the values of the thermodynamic functions for the activated complex and the **reactants** for a **chemical reaction** (all the substances being in their **standard states**).

heat of atomization Amount of heat that is required to convert 1 mole of an element into the gaseous state.

■ **heat of combustion** Heat change that accompanies the complete combustion of 1 mole of a substance.

■ **heat of formation** Heat change that occurs when 1 mole of a compound is formed from its elements, in their normal states.

■ **heat of neutralization** Heat change that occurs when 1 mole of aqueous **hydrogen ions** are neutralized by a **base** in dilute solution.

■ **heat of reaction** Heat change that occurs when the molar quantities (lowest possible multiples of 1 mole) of reactants as stated in a chemical equation react together.

■ **heat of solution** Heat change that occurs when 1 mole of a substance is dissolved in so much water that further dilution with water produces no further heat change.

■ **heavy oil** Alternative name for **gas oil**.

■ **heavy water** D_2O Compound of oxygen and **deuterium** (heavy hydrogen), analogous to **water**. It differs from ordinary water in most of its properties; *e.g.* it will not support life. Alternative name: deuterium oxide.

hecto- Metric prefix for a multiple of 10^2; denotes 100 times.

Heisenberg uncertainty principle The precise position and momentum of an electron cannot be determined simultaneously. It was named after the German physicist Werner Heisenberg (1901–76).

■ **helium** He Gaseous element in Group 0 of the Periodic Table (the **rare gases**), which occurs in some natural gas deposits. It is inert and non-inflammable, used to fill airships and balloons (in preference to inflammable hydrogen) and in helium-oxygen 'air' mixtures for divers (in preference to the nitrogen-oxygen mixture of real air, which can cause the bends). It is also used in gas lasers. Liquid helium is employed as a coolant in cryogenics. At. no. 2; r.a.m. 4.0026. [6/9/a]

Helmholtz free energy (*F*) Thermodynamic quantity equal to $U - TS$, where U is the internal energy, T the thermodynamic temperature and S the entropy. It was named after the German physicist Hermann Helmholtz (1821–94).

■ **hemihydrate** Compound containing one water molecule for every two molecules of the compound; *e.g.* $CaSO_4.\frac{1}{2}H_2O$.

Henry's law Weight of gas dissolved by a liquid is proportional to the gas pressure.

heparin Polysaccharide substance that prevents the clotting of blood by inhibiting the conversion of prothrombin to thrombin; used medicinally as an anticoagulant.

■ **heptane** C_7H_{16} Liquid **alkane** hydrocarbon, the seventh member of the methane series, present in petrol.

■ **heptavalent** With a valency of seven. Alternative name: septivalent.

Hess's law Total energy change resulting from a chemical reaction is dependent only on the initial and final states, and is independent of the reaction route. It was named after the Austrian physicist Victor Hess (1883–1964).

hetero- Prefix denoting other or different. *See also* **homo-**.

heterocyclic compound Cyclic organic compound that contains atoms other than carbon in the ring, *e.g.* **pyridine**. *See also* **homocyclic compound**.

■ **heterogeneous** Relating to more than one **phase**, *e.g.* describing a **chemical reaction** that involves one or more solids in addition to a gas or a liquid phase. *See also* **homogeneous**.

heterolytic fission Breaking of a two-electron **covalent bond** to give two fragments, with one fragment retaining both electrons. Alternative name: heterolytic cleavage. *See also* **homolytic fission**.

hexamine $C_6H_{12}N_4$ Organic compound made by condensing **formaldehyde** (methanal) with ammonia, used as a camping fuel and antiseptic drug; It can be nitrated to make the high explosive **cyclonite**. Alternative name: hexamethylenetetramine.

hexane C_6H_{14} Colourless liquid **alkane**, used as a **solvent**.

hexanedioic acid Alternative name for **adipic acid**.

hexose Monosaccharide carbohydrate (sugar) that contains six carbon atoms and has the general formula $C_6H_{12}O_6$; *e.g.* **glucose** and **fructose**.

high-vacuum distillation Alternative name for **molecular distillation**.

■ **histamine** Organic compound that is released from cells in connective tissue during an allergic reaction. It causes dilation of capillaries and constriction of bronchi.

histidine $(C_3H_3N_2)CH_2CH(NH_2)COOH$ Crystalline soluble solid, an optically active basic **essential amino acid**. Alternative name: 2-amino-3-imidazolylpropanoic acid.

histone One of a group of small **proteins** with a large proportion of basic **amino acids**; *e.g.* arginine or lysine. Histones are found in combination with nucleic acid in the chromatin of eukaryotic cells.

holmium Ho Silvery metallic element in Group IIIB of the Periodic Table (one of the **lanthanides**). At. no. 67; r.a.m. 164.930.

holoenzyme Enzyme that forms from the combination of a coenzyme and an apoenzyme. The former determines the nature and the latter the specificity of a reaction.

■ **homo-** Prefix denoting the same or similar. *See also* **hetero-**.

homocyclic compound Chemical compound that contains one or more closed rings comprising carbon atoms only, *e.g.* **benzene**. *See also* **heterocyclic compound**.

■ **homogeneous** Relating to a single **phase** (*e.g.* describing a **chemical reaction** in which all the reactants are solids, or liquids or gases); describing a system of uniform composition. *See also* **heterogeneous**.

■ **homologous series** Family of organic chemical compounds with the same general formula; *e.g.* **alkanes, alkenes** and **alkynes**.

■ **homologue** Member of a **homologous series**.

homolytic fission Breaking of a two-electron **covalent bond** in such a way that each fragment retains one electron of the bond. Alternative name: homolytic cleavage. *See also* **heterolytic fission**.

homopolymer **Polymer** that is formed by polymerization of a single substance (monomer); *e.g.* polyethylene, polypropylene. *See also* **copolymer**.

■ **hormone** *1.* In animals, chemical messenger of the body that is secreted directly into the bloodstream by an endocrine gland. Each gland secretes hormones of different composition and activities which exert specific effects on certain target tissues. *2.* In plants, organic substance that at very low concentrations affects growth and development; *e.g.* auxin, gibberellin, abscisin, florigen. [4/10/c]

hornblende Dark-coloured mineral, mainly composed of silicates of iron, calcium and magnesium, which is a major component of granite and many other metamorphic and igneous rocks.

■ **hydrate** Chemical compound that contains **water of crystallization**.

■ **hydrated** *1.* Describing a substance after treatment with water. *2.* Describing a compound that contains chemically bonded water, a **hydrate**.

■ **hydrated lime** Alternative name for **calcium hydroxide**.

■ **hydration** Attachment of water to the particles (particularly **ions**) of a **solute** during the dissolving process.

hydrazine NH_2NH_2 Colourless liquid, a powerful **reducing agent** used in organic synthesis and as a rocket fuel.

hydrazone Member of a family of organic compounds that contain the group $-C=NNH_2$. Hydrazones are formed by the action of **hydrazine** on an **aldehyde** or **ketone**, and are used in identifying them.

hydride Compound formed between **hydrogen** and another element (*e.g.* calcium hydride, CaH_2).

■ **hydro-** Prefix denoting water.

hydrobromic acid HBr Colourless acidic aqueous solution of **hydrogen bromide**; its salts are **bromides**.

■ **hydrocarbon** Organic compound that contains only carbon and hydrogen. The chief naturally occurring hydrocarbons are bitumen, coal, methane, natural gas and petroleum. Most of these, and hydrocarbons derived from them, are used as fuels. The aliphatic hydrocarbons form three homologous series: **alkanes**, **alkenes** and **alkynes**. Aromatic hydrocarbons are **cyclic** compounds.

■ **hydrochloric acid** HCl Colourless acidic aqueous solution of **hydrogen chloride**, a strong acid that dissolves most metals with the release of hydrogen; its salts are **chlorides**. It is used to make **chlorine** and other chemicals, and for cleaning metals before electroplating or galvanizing.

hydrocyanic acid HCN Very poisonous solution of **hydrogen cyanide** in water; its salts are **cyanides**. Alternative name: prussic acid.

hydrofluoric acid HF Colourless corrosive aqueous solution of **hydrogen fluoride**; its salts are **fluorides**. It is used for etching glass (and must be stored in plastic bottles).

■ **hydrogen** H Gaseous element usually given its own place at the beginning of the Periodic Table, but sometimes assigned to Group IA. Colourless, odourless and highly inflammable, it is the lightest gas known and occurs abundantly in combination in water (H_2O), coal and petroleum (mainly as hydrocarbons) and living things (mainly as carbohydrates). It is also the major constituent of the Sun and other stars, and is the most abundant element in the Universe. Hydrogen is made commercially by **electrolysis** of aqueous solutions, by **cracking** of petroleum, by the **Bosch process** or as **synthetic gas**. In the laboratory hydrogen is generally prepared by the action of a dilute acid on zinc. It has many uses: for hydrogenating (solidifying) oils, to make ammonia by the **Haber process**, as a fuel (particularly in rocketry), in organic synthesis, and as a **moderator** for nuclear reactors. In addition to the common form (sometimes called protium, r.a.m. 1.00797) there are two other isotopes: **deuterium** or heavy hydrogen (r.a.m. 2.01410) and the radioactive **tritium** (r.a.m. 3.0221). At. no. 1; r.a.m. (of the naturally occurring mixture of isotopes) 1.0080. [6/9/a]

■ **hydrogenation** Method of chemical synthesis by adding hydrogen to a substance. It forms the basis of many important industrial processes, such as the conversion of liquid oils to solid fats.

■ **hydrogenation of coal** Industrial synthesis of mineral oil from coal by **hydrogenation**.

■ **hydrogenation of oil** Manufacture of margarine by the **hydrogenation** of liquid vegetable oils to edible fats.

hydrogen bomb Powerful explosve device that uses the sudden release of energy from **nuclear fusion** (of **deuterium** and **tritium** atoms). Alternative name: thermonuclear bomb.

hydrogen bond Strong chemical **bond** that holds together some **molecules** that contain hydrogen, *e.g.* water molecules, which become associated as a result. A hydrogen atom bonded to an electronegative atom interacts with a (non-bonding) **lone pair of electrons** on another electronegative atom.

hydrogen bromide HBr Pale yellow gas which dissolves in water to form **hydrobromic acid**.

■ **hydrogencarbonate** Acidic salt containing the ion HCO_3^-. Alternative name: bicarbonate.

■ **hydrogen chloride** HCl Colourless gas which dissolves readily in water to form **hydrochloric acid**. It is made by treating a **chloride** with concentrated sulphuric acid or produced as a by-product of electrolytic processes involving chlorides.

hydrogen cyanide HCN Colourless poisonous gas, which dissolves in water to form **hydrocyanic acid** (prussic acid). It has a characteristic smell of bitter almonds.

hydrogen electrode Half-cell that consists of hydrogen gas bubbling around a platinum electrode, covered in platinum black (very finely divided platinum). It is immersed in a molar acid solution and used for determining standard **electrode potentials**. Alternative name: hydrogen half-cell.

hydrogen fluoride HF Colourless fuming liquid, which is extremely corrosive and dissolves in water to form **hydrofluoric acid**.

hydrogen half-cell Alternative name for **hydrogen electrode**.

■ **hydrogen halide** Compound of hydrogen and a **halogen**; *e.g.* hydrogen fluoride, HF, hydrogen iodide, HI.

■ **hydrogen ion** H^+ Positively-charged hydrogen atom; a **proton**. A characteristic of an **acid** is the production of hydrogen ions, which in aqueous solution are hydrated to hydroxonium ions, H_3O^+.

■ **hydrogen peroxide** H_2O_2 Colourless syrupy liquid with strong oxidizing powers, soluble in water in all proportions. Dilute solutions are used as an **oxidizing agent**, disinfectant and bleach; in concentrated form it is employed as a rocket fuel.

■ **hydrogensulphate** Acidic salt containing the ion HSO_4^-. Alternative name: bisulphate.

■ **hydrogen sulphide** H_2S Colourless poisonous gas with a characteristic smell (when impure) of bad eggs. It is formed by rotting sulphur-containing organic matter and the action of acids on sulphides.

■ **hydrogensulphite** Acidic salt containing the ion HSO_3^-. Alternative name: bisulphite.

hydrolysis Chemical decomposition of a substance by water, with a hydroxyl group (–OH) from the water taking part in the reaction; *e.g.* esters hydrolyse to form alcohols and acids.

hydrophilic Possessing an affinity for water.

hydrophobic Water-repellent; having no attraction for water.

hydrosol Aqueous solution of a **colloid**.

hydrous Containing water.

hydroxide Compound of a metal that contains the **hydroxyl group** (–OH) or the hydroxide ion (OH^-). Many metal hydroxides arc **bases**.

hydroxonium ion Hydrated **hydrogen ion**, H_3O^+.

hydroxybenzene Alternative name for **phenol**.

hydroxybenzoic acid Alternative name for **salicylic acid**.

hydroxyl group (–OH) Group containing oxygen and hydrogen, characteristics of **alcohols** and some **hydroxides**.

hydroxypropionic acid Alternative name for **lactic acid**.

hygroscopic Having the tendency to absorb moisture from the atmosphere.

hyperon Member of a group of short-lived **elementary particles** which are greater in mass than the **neutron**.

hypo Popular name for **sodium thiosulphate**.

hypochlorite Salt of **hypochlorous acid** (containing the ion ClO^-). Hypochlorites are used as bleaches and disinfectants.

hypochlorous acid (HOCl) Weak liquid **acid** stable only in solution, used as an oxidizing agent and bleach. Its salts are hypochlorites.

hyposulphuric acid Alternative name for **dithionic acid**.

I

■ **ice** Water in its solid state (*i.e.* below its **freezing point**, 0°C). It is less dense than water, because **hydrogen bonds** give its crystals an open structure, and it therefore floats on water. This also means that water expands on freezing.

■ **Iceland spar** Very pure transparent form of **calcite** (calcium carbonate), and noted for the property of double refraction.

■ **ideal crystal** Crystal structure that is considered as perfect, containing no **defects**.

ideal gas Hypothetical gas with molecules of negligible size that experience no intermolecular forces. Such a gas would in theory obey the **gas laws** exactly. Alternative name: perfect gas. *See also* **real gas**.

ideal solution Hypothetical solution that obeys **Raoult's law** exactly.

ignis fatuus Light seen over marshy ground caused by the ignition of **methane** (generated in rotting vegetation) by traces of phosphorus compounds, *e.g.* phosgene. Alternative name: will-o'- the-wisp.

■ **ignition** Initial combustion of a substance, particularly of an explosive mixture (*e.g.* of petrol vapour and air) in an internal combustion engine.

■ **ignition temperature** Temperature to which a substance must be heated before it will burn in air (or in some other specified oxidant).

imide Organic compound derived from an acid anhydride, general formula R-CONHCO-R′, where R and R′ are organic **radicals**. Alternative name: imido compound.

imido compound Alternative name for **imide.**

imido-urea Alternative name for **guanidine.**

imine Secondary **amine**, an organic compound derived from **ammonia**, general formula RNHR′, where R and R′ are organic **radicals**; *e.g.* dimethylamine, $(CH_3)NH$. Alternative name: imino compound.

imino compound Alternative name for **imine.**

imino group The group =NH, characteristic of an **imine.**

■ **immiscible** Describing two or more liquids that will not mix (when shaken together they separate into layers).

■ **indicator** Substance that changes colour to indicate the end of a chemical reaction or the **pH** of a solution; *e.g.* litmus, methyl orange, phenolphthalein. Indicators are commonly used in titrations in **volumetric analysis.** [6/5/b]

indigo $C_{16}H_{10}N_2O_2$ Blue organic dye, a derivative of **indole,** that occurs as a glucoside in plants of the genus *Indigofera.*

indium In Silvery-white metallic element of Group IIIA of the Periodic Table, used in making mirrors and **semiconductors.** At. no. 49; r.a.m. 114.82.

indole C_8H_7N Colourless solid, consisting of fused **benzene** and **pyrrole** rings, which occurs in coal-tar and certain plants. It is the basis of the **indigo** molecule. Alternative name: benzpyrrole.

■ **induced radioactivity** Alternative name for **artificial radioactivity.**

■ **inert** Chemically nonreactive, *e.g.* gold is inert in air at normal temperatures.

■ **inert gas** Member of Group 0 of the Periodic Table; the unreactive elements **helium, neon, argon, krypton, xenon** and **radon**. Alternative name: rare gas. [6/9/a]

■ **inorganic chemistry** Study of chemical elements and their **inorganic compounds**. *See also* **organic chemistry**.

■ **inorganic compound** Compound that does not contain **carbon**, with the exception of carbon's oxides, metallic carbides, carbonates and hydrogencarbonates. *See also* **organic compounds**.

■ **insoluble** Describing a substance that does not dissolve in a given **solvent**; not capable of forming a **solution**.

■ **insulator** Substance that is a poor conductor of electricity or heat; a non-conductor. Most non-metallic elements (except carbon) and **polymers** are good insulators. [11/3/a]

interface Boundary of contact (the common surface) of two adjacent **phases**, either or both of which may be solid, liquid or gaseous.

intermediate *1*. In industrial chemistry, compound to be subjected to further chemical treatment to produce finished products such as dyes and pharmaceuticals. *2*. Short-lived species in a complex chemical reaction.

intermediate compound Compound of two or more metals that are present in definite proportion although they frequently do not follow normal **valence** rules.

intermediate neutron Neutron that has energy between that of a **fast neutron** and a **thermal neutron**.

intermolecular force Force that binds one molecule to another. Intermolecular forces are much weaker than the bonding forces that hold together the atoms of a molecule. *See also* **van der Waals' force**.

internal conversion Effect on the nucleus of an atom produced by a gamma-ray photon emerging from it and giving up its energy on meeting an electron of the same atom.

interstitial Describing an atom that is in a position other than a normal **lattice** place.

inversion In organic chemistry, splitting of **dextrorotatory** higher sugars (*e.g.* sucrose) into equivalent amounts of **laevorotatory** lower sugars (*e.g.* fructose and glucose).

invertase Plant **enzyme** that brings about the hydrolysis of **sucrose** (cane-sugar) to form **invert sugar**.

invert sugar Natural **disaccharide** sugar that consists of a mixture of **fructose** and **glucose**, found in many fruits. It is also formed by the hydrolysis of **sucrose** (cane-sugar), which can be brought about by the enzyme **invertase**.

iodide Compound of **iodine** and another element; salt of hydriodic acid (HI).

iodine I Non-metallic element in Group VIIA of the Periodic Table (the **halogens**), extracted from Chile saltpetre (in which it occurs as an iodate impurity of sodium nitrate) and certain seaweeds. It forms purple-black crystals that sublime on heating to produce a violet vapour. It is essential for the correct functioning of the thyroid gland. Iodine and its organic compounds are used in medicine; silver iodide is used in photography. At. no. 53; r.a.m. 126.9044. [6/9/a]

iodine number Number that indicates the amount of iodine taken up by a substance, *e.g.* by fats or oils; it gives a measure of the number of **unsaturated** bonds present. Alternative names: iodine value, iodine absorption.

iodoform Alternative name for **triiodomethane**.

■ **ion** Atom or molecule that has positive or negative electric charge because of the loss or gain of one or more electrons. Many inorganic compounds dissociate into ions when they dissolve in water. Ions are the electric current-carriers in **electrolysis** and in discharge tubes. [8/10/b]

■ **ion exchange** Reaction in which **ions** of a solution are retained by oppositely-charged groups covalently bonded to a solid support, such as **zeolite** or a synthetic **resin**. The process is used in water softeners, **desalination** plants and for **isotope** separation. [5/6/a]

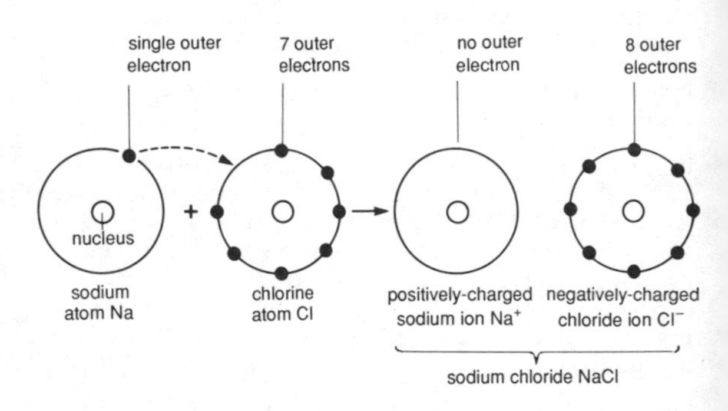

Formation of an ionic bond

■ **ionic bond** Chemical bond that results from the attractive electrostatic force between oppositely charged **ions** in molecules or ionic crystals. Alternative name: electrovalent bond. *See also* **covalent bond**. [8/8/c]

■ **ionic crystal** Crystal composed of **ions**. Alternative names: electrovalent crystal, polar crystal.

■ **ionic product** Product (in moles per litre) of the concentrations of the **ions** in a liquid or solution, *e.g.* in sodium chloride solution, the ionic product of sodium chloride is given by $[Na^+][Cl^-]$. In a pure liquid, it results from the **dissociation** of molecules in the liquid.

ionic radius Radius of an **ion** in a crystal.

■ **ionization** Formation of **ions**. It is generally achieved by chemical or electrical processes, or by **dissociation** of ionic compounds in solution, although at extremely high temperatures (such as those in stars) heat can cause ionization.

ionization potential Electron bonding energy, the energy required to remove an **electron** from a neutral atom.

■ **ionizing radiation** Any **radiation** that causes **ionization** by producing **ion pairs** in the medium it passes through. [8/9/b]

ion pair Two charged fragments that result from simultaneous ionization of two uncharged ones; a positive and a negative **ion**.

ion pump High-vacuum pump for removing a gas from a system by ionizing its atoms or molecules and adsorbing the resulting **ions** on a surface.

iridium Ir Steel-grey metallic element in Group VIII of the Periodic Table (a **transition element**). It is used (with platinum or osmium) in hard alloys for bearings, surgical tools and crucibles. At. no. 77; r.a.m. 192.2.

■ **iron** Fe Silver-grey magnetic metallic element in Group VIII of the Periodic Table (a **transition element**). It is the fourth most abundant element in the Earth's crust and probably forms much of the core. It is also the most widely used metal, particularly (alloyed with carbon and other elements) in **steel**. It occurs in various ores, chief of which are **haematite, limonite** and **magnesite,** which are refined in a **blast furnace** to produce **pig iron**. Inorganic iron compounds are used as pigments; the blood pigment **haemoglobin** is an organic iron compound. At. no. 26; r.a.m. 55.847.

■ **iron(II)** Alternative name for ferrous in iron compounds.

■ **iron(III)** Alternative name for ferric in iron compounds.

iron(III) chloride $FeCl_3.6H_2O$ Brown crystalline compound, used as a catalyst, a **mordant** and for etching copper in the manufacture of printed circuits. Alternative name: ferric chloride.

■ **iron(III) oxide** Fe_2O_3. Red insoluble compound, the principal constituent of **haematite**. It is used as a pigment, catalyst and polishing compound. Alternative name: ferric oxide.

iron(II) sulphate $FeSO_4.7H_2O$. Green crystalline compound, used in making inks, in printing and as a wood preservative. Alternative name: ferrous sulphate.

■ **irreversible reaction** Chemical reaction that takes place in one direction only, therefore proceeding to completion.

isocyanide Organic compound of general formula RNC, where R is an organic **radical**; *e.g.* methyl isocyanide, CH_3NC. Alternative names: isonitrile, carbylamine.

isoelectric point Hydrogen ion concentration **(pH)** at which a system is electrically neutral.

isoleucine Crystalline **amino acid**; it is a constituent of **proteins**, and essential in the diet of human beings.

■ **isomer** Substance that exhibits chemical **isomerism**.

■ **isomerism** Existence of substances that have the same molecular composition, but different structures. *See* **optical isomerism**.

■ **isomorphism** Phenomenon in which two or more minerals that are closely similar in chemical composition crystallize in the same forms.

isonitrile Alternative name for **isocyanide**.

isoprene $CH_2=CH(CH_3)CH=CH_2$ Colourless unsaturated liquid hydrocarbon, used in making synthetic rubber. Alternative name: 2-methyl-1,3-butadiene.

isopropanol $(CH_3)_2CHOH$ One of the two **isomers** of **propanol**. Alternative name: isopropyl alcohol.

isopropyl alcohol Alternative name for **isopropanol**.

■ **isotope** One form of an atom that has the same **atomic number** but a different **atomic mass** to other forms of that atom. This results from there being different numbers of **neutrons** in the nuclei of the atoms, *e.g.* uranium-238 (also written as U-238 or $_{238}U$) and uranium-235 are two isotopes of uranium with atomic masses of 238 and 235 respectively. The isotopes of an atom are chemically identical, although with very light isotopes the relative difference in masses may make them react at different rates. The existence of isotopes explains why most elements have non-integral atomic masses. A few elements have no naturally-occurring isotopes, including fluorine, gold, iodine and phosphorus.

- **isotopic number** Difference between the number of **neutrons** in an **isotope** and the number of **protons**. Alternative name: neutron excess.
- **isotopic weight** **Atomic mass** of an **isotope**. Alternative name: isotopic mass.

K

kainite Hydrated **magnesium sulphate** mineral containing **potassium chloride**; it crystallizes in the **monoclinic** system.

■ **kaolin** China clay, a mineral consisting of **kaolinite**. [9/3/d]

kaolinite Hydrated aluminium silicate mineral, consisting of minute crystals derived from **feldspars**.

Kekulé structure Structural forms of **benzene** with alternate single and double bonds, proposed by the German chemist Friedrich August Kekulé (1829–96).

■ **kelvin** (K) SI unit of thermodynamic temperature, named after the British physicist Lord Kelvin (William Thomson) (1824–1907). It is equal in magnitude to a degree Celsius (°C). *See also* **Kelvin temperature**. [13/3/b]

■ **Kelvin temperature** Scale of temperature that originates at **absolute zero**, with the **triple point** of water defined as 273.16K. The freezing point of water (on which the Celsius scale is based) is 273.16K. Alternative name: Kelvin thermodynamic scale of temperature. [13/3/b]

■ **kerosene** Oily mixture of **hydrocarbons**, obtained mainly from **petroleum** and **oil shale**. It has a boiling range from about 150°C to 300°C, and is used as a fuel and as a solvent. Alternative names: kerosine, paraffin oil.

ketene Member of a group of unstable organic compounds of general formula $R_2C=CO$, where R is hydrogen or an organic **radical**; *e.g.* ketene, $CH_2=CO$. Ketenes react with other **unsaturated** compounds to form 4-membered rings.

keto-enol tautomerism Existence of a chemical compound in two double-bonded structural forms, keto and enol, which are in equilibrium. The keto form (containing the group $-CH_2-C=O-$) changes to the enol (containing $-CH=C(OH)-$) by the migration of a hydrogen atom to form a **hydroxyl group** with the oxygen of the **carbonyl group**; the position of the double bond also changes.

ketone Member of a family of organic compounds of general formula RCOR′, where R and R′ are organic **radicals** and $=CO$ is a **carbonyl group**. Ketones may be made in various ways, such as the oxidation of a secondary **alcohol**; *e.g.* oxidation of **isopropanol**, $(CH_3)_2OH$, gives **acetone** (propanone), $(CH_3)_2CO$.

keV Abbreviation of kilo-electron-volt, a unit of particle energy equivalent to 10^3 electron-volts.

kieselguhr Silica-containing mineral, a whitish powder that consists mainly of diatom skeletons. It is used in fireproof cements and in the manufacture of **dynamite**. Alternative name: diatomite.

■ **kilo-** Metric prefix meaning a thousand times ($x10^3$).

■ **kilogram** (kg) SI unit of mass, equal to 1,000 grams. 1 kg = 2.2046 lb.

kiloton Unit for the power of a nuclear explosion or warhead, equivalent to 1,000 tonnes of TNT.

kinetics Study of the rates at which chemical reactions take place.

■ **kinetic theory** Theory that accounts for the properties of substances in terms of the movement of their component particles (atoms or molecules). The theory is most important in describing the behaviour of gases (when it is referred to as the kinetic theory of gases). An ideal gas is assumed to be

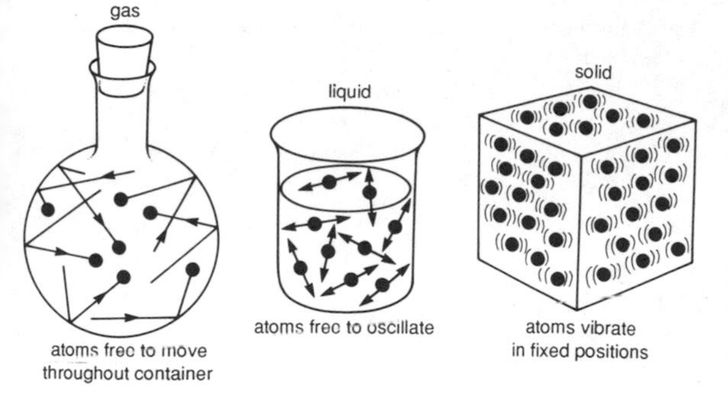

Kinetic theory accounts for how substances behave

made of perfectly elastic particles that collide only occasionally with each other. Thus, *e.g.*, the pressure exerted by a gas on its container is then the result of gas particles colliding with the walls of the container. [8/7/a]

■ **knock** Pre-ignition, the instantaneous violent explosion of a compressed mixture of fuel vapour and air in an internal combustion engine. It reduces the available mechanical energy, but can be prevented by adding an anti-knock agent such as **tetraethyllead** to the fuel. Alternative names: knocking, pinking.

krypton Kr Gaseous nonmetallic element of Group 0 of the Periodic Table (the **rare gases**), which occurs in trace quantities in air (from which it is extracted). It is used in gas-filled lamps and discharge tubes. At. no. 36; r.a.m. 83.80.

kurchatovium Alternative name for the **post-actinide** element **rutherfordium**.

L

■ **labelling** Technique in which an atom (often in a molecule of a compound) is replaced by its radioactive **isotope** (termed a tracer) as a means of locating the compound or following its progress; *e.g.* in a plant or animal, or in a chemical reaction. [8/7/d]

labile Unstable; usually applied with respect to particular conditions, *e.g.* heat-labile.

lactate Salt or ester of **lactic acid** (2– hydroxypropanoic acid).

lactic acid $CH_3CH(OH)COOH$ Colourless liquid organic acid. A mixture of (+)-lactic acid (**dextrorotatory**) and (–)-lactic acid (**laevorotatory**) is produced by bacterial action on the sugar **lactose** in milk during souring. The (+)-form is produced in animals when **anaerobic respiration** takes place in muscles because of an insufficient oxygen supply during vigorous activity. Lactic acid is used in the chemical and textile industries. Alternative name: 2-hydroxypropanoic acid.

lactoflavin Alternative name for **riboflavin**.

lactone Unstable internal ester, a **cyclic** compound containing the group –CO.O– in the ring, formed by heating γ- and δ-hydroxycarboxylic acids.

■ **lactose** $C_{12}H_{22}O_{11}$ White crystalline **disaccharide** sugar that occurs in milk, formed from the union of **glucose** and **galactose**. It is a reducing sugar. Alternative name: milk-sugar.

laevorotatory Describing a compound with **optical activity** that causes the plane of polarized light to rotate in an anti-clockwise direction. Indicated by the prefix (–)- or *l*-.

laevulose Alternative name for **fructose**.

lambda particle Type of **elementary particle** with no electric charge.

■ **lamp black** Type of **carbon black**.

■ **lanolin** Yellowish sticky substance obtained from the grease that occurs naturally in wool. It is used in cosmetics, as an ointment and in treating leather. Alternative names: lanoline, wool fat.

lanthanide Member of the Group IIIB elements of atomic number 57 to 71. The properties of these metals are very similar, and consequently they are difficult to separate. Alternative names: lanthanoid, rare-earth element.

lanthanum La Silver-white metallic element in Group IIB of the Periodic Table, the parent element of the **lanthanide** series. It is used in making lighter-flints. At. no. 57; r.a.m. 138.91.

lapis lazuli Semi-precious deep blue gem, a sodium aluminium silicate that contains sulphur.

Larmor precession Orbital motion of an **electron** about the **nucleus** of an atom when it is subjected to a small magnetic field. The electron precesses about the direction of the magnetic field. It was named after the British physicist Joseph Larmor (1857–1942).

laterite Fine-grained clay produced by the weathering of igneous rocks in a tropical climate. The presence of iron(III) hydroxide gives it a distinctive red colour.

■ **latex** Milky fluid produced in some plants after damage, containing sugars, proteins and alkaloids. It is used in manufacture, *e.g.* of rubber. A suspension of synthetic rubber is also called latex.

■ **lattice** Regular network of atoms, ions or molecules in a **crystal**.

lattice energy Strength of an **ionic bond**; the energy required for the separation of the ions in 1 mole of a crystal to an infinite distance from each other. Alternative name: lattice enthalpy.

■ **laughing gas** Alternative name for **dinitrogen oxide**.

lauric acid $CH_3(CH_2)_{10}COOH$ White crystalline **carboxylic acid**, used in making **soaps** and **detergents**. Alternative names: dodecanoic acid, dodecylic acid.

lawrencium Lr Radioactive element in Group IIIB of the Periodic Table (one of the **actinides**), the heaviest element definitely identified. At. no. 103; r.a.m. 257 (most stable isotope).

LD – 50 (lethal dose 50) Toxicity test in which the end-point is the quantity of a substance that causes death in 50% of the organisms tested.

■ **leaching** Washing out of a soluble material from a solid by a suitable liquid.

■ **lead** Pb Silver-blue poisonous metallic element in Group IVA of the Periodic Table, obtained mainly from its sulphide ore **galena**. Various **isotopes** of lead are final elements in **radioactive decay** series. The metal is used in building, as shielding against ionizing **radiation**, as electrodes in **lead-acid accumulators** and in various alloys (such as **solder**, metals for bearings and type metal). Its inorganic compounds are used as pigments; tetraethyllead is employed as an anti-knock agent in petrol. At. no. 82; r.a.m. 207.19.

lead(II) Alternative name for plumbous in lead compounds.

lead(IV) Alternative name for plumbic in lead compounds.

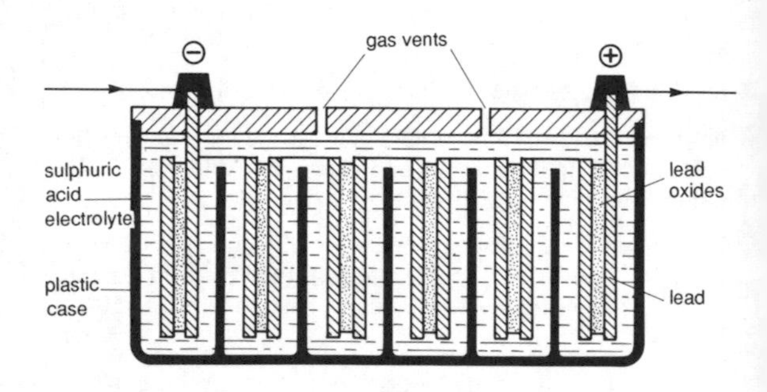

Construction of a lead-acid accumulator

■ **lead-acid accumulator** Rechargable electrolytic cell (battery) that has positive electrodes of lead(IV) oxide (PbO_2), negative electrodes of lead, and a solution of **sulphuric acid** as the electrolyte.

■ **lead-chamber process** Obsolete process for the manufacture of **sulphuric acid** that used sulphur dioxide, air and nitrogen oxides as a catalyst. Alternative name: chamber process. *See also* **contact process.**

lead dioxide Alternative name for **lead(IV) oxide**.

lead equivalent Factor that compares any form of shielding against radioactivity to the thickness of lead that would provide the same measure of protection.

lead-free petrol Petrol that does not contain the anti-knocking agent **tetraethyllead** (which causes atmospheric pollution).

lead monoxide Alternative name for **lead(II) oxide**.

lead(II) oxide PbO Yellow crystalline substance, used in the manufacture of glass. Alternative names: lead monoxide, litharge.

lead(IV) oxide PbO_2 Brown amorphous solid, a strong oxidizing agent, used in lead-acid accumulators. Alternative names: lead dioxide, lead peroxide.

lead tetraethyl(IV) Alternative name for **tetraethyllead**.

■ **Le Chatelier's principle** If a change occurs in one of the factors (such as temperature or pressure) under which a system is in equilibrium, the system will tend to adjust itself so as to counteract the effect of that change. It was named after the French physicist Henri le Chatelier (1850–1936). Alternative name: Le Chatelier-Braun principle. [7/8/a]

■ **Leclanché cell Primary cell** that has a zinc cathode and carbon anode dipping into an electrolyte of ammonium chloride solution. A porous pot of crushed carbon and manganese(IV) oxide (manganese dioxide) surrounds the anode to prevent **polarization**. It is the basis of the **dry cell** used in most batteries. It was named after the French chemist Georges Leclanché (1839–82).

lepton Subatomic particle that does not interact strongly with other particles, *e.g.* an electron. *See also* **hadron**.

leucine $(CH_3)_2CHCH_2CH(NH_2)COOH$ Colourless crystalline **amino acid**; a constituent of many **proteins**. Alternative name: 2-amino-4-methylpentanoic acid.

Lewis acid and base Concept of **acids** and **bases** in which an acid is defined as a substance capable of accepting a pair of electrons, whereas a base is able to donate a pair of electrons to a bond. It was named after the American chemist Gilbert Lewis (1875–1946).

ligand Molecule or **ion** that has at least one electron pair that donates its electrons to a metal ion or other electron acceptor, often forming a **co-ordinate bond**.

ligase Enzyme that repairs damage to the strands that make up **DNA**, widely used in recombination techniques to seal the joins between DNA sequences.

■ **light petroleum ether** Mixture of **pentane** and **hexane** hydrocarbons derived from crude petroleum; boiling range 30 to 60°C. It is used as a solvent.

lignin Complex **polymer** found in many plant cell walls, that glues together fibres of **cellulose** and provides additional support for the cell wall.

lignite Soft brown form of **coal**, between peat and bituminous coals in quality. It generally has a high moisture content and when burnt gives only about half as much heat as good-quality coal. Alternative name: brown coal.

ligroin Mixture of hydrocarbons derived from crude petroleum, boiling range 80 to 120°C, used as a general solvent.

■ **lime** *1.* General term for quicklime (**calcium oxide**, CaO), slaked lime and hydrated lime (both **calcium hydroxide**, $Ca(OH)_2$). They are obtained from **limestone**. *2.* Ground limestone used as a fertilizer and in iron smelting.

■ **limestone** Sedimentary rock of marine origin. Its main constituent is **calcium carbonate**, and it is used as a building stone, in iron smelting and in the manufacture of **lime**. [9/3/d]

■ **lime water** Solution of **calcium hydroxide** ($Ca(OH)_2$) in water, used as a test for carbon dioxide (which turns lime water milky when bubbled through it due to the precipitation of calcium carbonate; after prolonged bubbling the solution goes clear again due to the formation of soluble calcium hydrogencarbonate).

■ **limonite** Major iron ore that consists of oxides and hydroxide of **iron**. [9/3/d]

Linde process Technique for liquefying air, and extracting liquid oxygen and liquid nitrogen. It was named after the German scientist Carl von Linde (1842–1934).

linear molecule Molecule whose atoms are arranged in a line.

linoleic acid $C_{17}H_{31}COOH$ **Unsaturated fatty acid**, used in paints, which occurs in **linseed oil** and other oils derived from plants. Alternative name: linolic acid.

■ **linseed oil** Oil extracted from the seeds of flax. Because it contains **linoleic acid**, it is a drying oil (hardening on exposure to air), used in paints, putty, varnishes and enamels.

lipase In vertebrates, an **enzyme** in intestinal juice and pancreatic juice that catalyses the **hydrolysis** of **fats** to **glycerol** and **fatty acids**.

lipid Member of a group of naturally occurring fatty or oily compounds that share the property of being soluble in organic solvents, but sparingly soluble in water. Also, all lipids yield **monocarboxylic acids** on **hydrolysis**.

lipochrome Yellow pigment in butterfat.

■ **liquefaction of gases** All gases can be liquefied by a

combination of cooling and compression. The greater the pressure, the less the gas needs to be cooled, but there is for each gas a certain critical temperature below which it must be cooled before it can be liquefied.

■ **liquefied natural gas** (LNG) Liquid **methane**. *See* **natural gas**.

■ **liquefied petroleum gas** (LPG) Mixture of hydrocarbons derived from crude petroleum which has **propane** as its major constituent. It is used as a fuel.

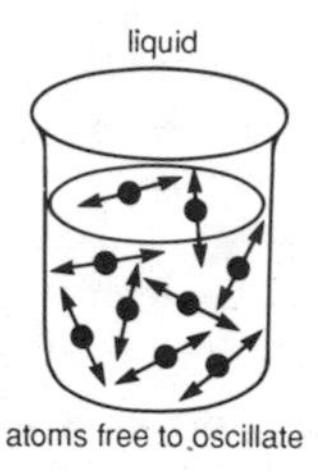

Molecules in a liquid are free to move around

■ **liquid** Fluid that, without changing its volume, takes the shape of its container. According to the **kinetic theory**, the molecules in a liquid are not bound together as rigidly as those in a solid but neither are they as free to move as those of a gas. It is therefore a **phase** that is intermediate between a solid and a gas.

■ **liquid crystal** Compound that is liquid at room temperature and atmospheric pressure but shows characteristics normally expected only from solid crystalline substances. Large groups of its molecules maintain their mobility but nevertheless also retain a form of structural relationship. Some liquid crystals change colour according to the temperature.

liquid-liquid extraction Alternative name for **solvent extraction**.

■ **litharge** Alternative name for **lead(II) oxide**.

lithium Li Silver-white metallic element in Group IA of the Periodic Table (the **alkali metals**), the solid with the least density. Its compounds are used in lubricants, ceramics, drugs and the plastics industry. At. no. 3; r.a.m. 6.939.

lithium aluminium hydride $LiAlH_4$ Powerful reducing agent, used in organic chemistry. Alternative name: lithium tetrahydridoaluminate(III).

■ **litmus** Dye made from certain lichens, used as an **indicator** to distinguish acids from alkalis. Neutral litmus solution or litmus paper is naturally violet-blue; acids turn it red, alkalis turn it blue. [6/5/b]

■ **litre** (l) Unit of volume in the metric system, defined as 1 dm^3, *i.e.* 1,000 cm^3 (formerly defined as the volume of 1 kg of water at 4°C). 1 litre = 1.7598 pints.

■ **LNG** Abbreviation of **liquefied natural gas**.

■ **lodestone** Fe_3O_4 Naturally occurring magnetic oxide of **iron**. Alternative names: loadstone, magnetite. [9/3/d]

lone pair of electrons Pair of unshared electrons of opposite spin (in the same **orbital**) that under suitable conditions can form a **co-ordinate bond**. *E.g.* the nitrogen atom in ammonia

(NH_3) has a lone pair of electrons, which form bonds in various co-ordination compounds.

low-energy electron diffraction **Electron diffraction** that uses low-energy electrons, which are strongly diffracted by surface layers of atoms.

Lowry-Brönsted theory *See* **Brönsted-Lowry theory**.

■ **LPG** Abbreviation of **liquefied petroleum gases**.

■ **LSD** Abbreviation of **lysergic acid diethylamide**. [3/6/e]

lutetium Metallic element in Group IIIB of the Periodic Table (one of the **lanthanides**). The irradiated metal is a beta-particle emitter, used in catalytic processes. At. no. 71; r.a.m. 174.97.

lyddite Explosive consisting of **picric acid**.

lyophilic Possessing an affinity for liquids.

lyophobic Liquid-repellent, having no attraction for liquids.

■ **lysergic acid diethylamide** (LSD) Synthetic substance, similar to some fungus **alkaloids**, which provokes hallucinations and extreme mental disturbance if taken, even in extremely small quantities. [3/6/e]

lysine $H_2N(CH_2)_4CH(NH_2)COOH$. **Essential amino acid** that occurs in **proteins** and is responsible for their basic properties because of its two $-NH_2$ groups. Alternative name: diaminocaproic acid.

lysozyme **Enzyme** in saliva, egg white, tears and mucus. It catalyses the destruction of bacterial cell walls by **hydrolysis**, and thus has a bactericidal effect.

M

■ **macro-** Prefix meaning large or long in size or duration.

macrocyclic Describing a chemical compound whose molecules have a large ring structure.

macromolecular Describing the structure of a chemical compound that consists of **macromolecules**.

■ **macromolecule** Very large **molecule** containing hundreds or thousands of **atoms**; *e.g.* natural **polymers** such as cellulose, rubber and starch, and synthetic ones, including **plastics**.

■ **magnesia Magnesium oxide**, MgO, particularly a form that has been processed and purified. It is used as an antacid.

■ **magnesite** Naturally occurring **magnesium carbonate**, $MgCO_3$, used as a refractory for furnace linings and to make various grades of **magnesium oxide** (magnesia).

■ **magnesium** Mg Reactive silver-white metallic element in Group IIA of the Periodic Table (the **alkaline earths**). It burns in air with a brilliant white light, and is used in flares and lightweight alloys. It is the metal atom in chorophyll and an important trace element in plants and animals. At. no. 12; r.a.m. 24.305.

magnesium carbonate $MgCO_3$ White crystalline compound, soluble in acids and insoluble in water and alcohol, which occurs naturally as **magnesite** and **dolomite**. It is used, often as the basic carbonate, as an antacid.

magnesium chloride $MgCl_2$ White crystalline compound obtained from sea-water and the mineral **carnallite**, used as a source of magnesium. The hexahydrate is hygroscopic and used as a moisturizer for cotton in spinning.

magnesium hydroxide $Mg(OH)_2$ White crystalline compound, used as an antacid.

magnesium oxide MgO White crystalline compound, insoluble in water and alcohol, made by heating **magnesium carbonate**. It is used as a refractory and antacid. Alternative name: magnesia.

magnesium silicate *See* **talc**.

■ **magnesium sulphate** $MgSO_4.7H_2O$ White crystalline salt, used in mineral waters, as a laxative and in the leather industry. Alternative name: Epsom salt.

■ **magnetite** Mineral form of a black iron oxide (Fe_3O_4), used as a source of iron, as a flux and in ceramics. It is strongly magnetic (*see* **lodestone**). [9/3/d]

■ **malachite** Bright green mineral consisting of a mixture of copper(II) carbonate and copper(II) hydroxide, $CuCO_3.CU(OH)_2$.

■ **malic acid** $COOHCH_2CH(OH)COOH$ Colourless crystalline **dicarboxylic acid** with an agreeable sour taste resembling that of apples, found in unripe fruit. It is used as a flavouring agent.

■ **malleable** Describing a metal that can easily be beaten into a sheet (*e.g.* lead, gold).

Mallory cell Alternative name for **mercury cell**.

■ **maltose** $C_{12}H_{22}O_{11}$ Common **disaccharide** sugar, composed of two molecules of **glucose**. It is found in **starch** and **glycogen**, and used in the food and brewing industries. [7/5/b]

manganese Mn Metallic element in group VIIA of the Periodic Table, used mainly for making special alloy steels and as a deoxidizing agent. It is also an essential trace element for plants and animals. At. no. 25, r.a.m. 54.94.

manganese dioxide Alternative name for **manganese(IV) oxide**.

■ **manganese(IV) oxide** MnO_2 Black amorphous compound, used as an oxidizing agent, catalyst and depolarizing agent in dry batteries. Alternative name: manganese dioxide.

manganic Alternative name for manganese(III) in manganese compounds.

manganin Alloy of copper with manganese and nickel that exhibits high electrical resistance, and which is used in resistors.

manganous Alternative name for manganese(II) in manganese compounds.

mannitol $HO.CH_2(CHOH)_4CH_2OH$ Soluble hexahydric alcohol that occurs in many plants, used as a sweetener and in medicine as a diuretic.

marble Metamorphic rock consisting of **calcium carbonate** ($CaCO_3$), derived from **limestone**.

■ **mass** Quantity of matter in an object, and a measure of the extent to which it resists acceleration if acted on by a force (*i.e.* its inertia). The SI unit of mass is the kilogram.

mass action, law of The driving force of a homogeneous chemical reaction is proportional to the active masses of the reacting substances.

mass decrement *See* **mass defect**.

mass defect *1.* Difference between the mass of an atomic **nucleus** and the masses of the particles that make it up, equivalent to the **binding energy** of the nucleus (expressed in mass units). *2.* Mass of an **isotope** minus its mass number. Alternative names: mass decrement, mass excess.

■ **mass number** (A) Total number of **protons** and **neutrons** in an atomic **nucleus**. Alternative name: nucleon number. *See also* **isotope**. [8/8/d]

mass spectrograph Vacuum system in which positive rays of charged atoms (**ions**) are passed through electric and magnetic fields so as to separate them in order of their charge-to-mass ratios on a photographic plate. It measures atomic masses of **isotopes** with precision.

mass spectrometer Mass spectrograph that uses electrical methods rather than photographic ones to detect charged particles.

mass spectrum Indication of the distribution in mass, or in mass-to-charge ratio, of ionized atoms or molecules produced by a **mass spectrograph**.

■ **matter** Substance that occupies space and has the property of inertia. These two characteristics distinguish matter from energy, the various forms of which make up the rest of the material Universe. *See also* **mass**.

Maxwell-Boltzmann distribution Law that describes the distribution of energy among molecules of a gas in thermal equilibrium.

mean free path Average distance that a gas molecule moves between two successive collisions with other molecules.

mean free time Average time between collisions of *1.* gas molecules; *2.* electrons and impurity atoms in a **semiconductor**.

mean life *1.* Average time for which the unstable **nucleus** of a radioisotope exists before decaying. *2.* Average time of survival of an elementary particle, ion, etc., in a given medium or of a charge carrier in a **semiconductor**.

meerschaum Very fine-grained claylike mineral composed of hydrated magnesium silicate, used for making smoking pipes and ornamental sculptures.

■ **mega-** Metric prefix meaning million times; $\times 10^6$ (*e.g.* megaton).

megaton Measure of the explosive power of a nuclear explosion or warhead, equivalent to a million tonnes of TNT.

■ **melamine** $C_3N_3(NH_2)_3$ White solid organic compound which consists of three amino groups ($-NH_2$) attached to a six-membered **heterocyclic** ring of alternate carbon and nitrogen atoms. It condenses with **formaldehyde** (methanal) or other **aldehydes** to form artificial **resins** that have excellent resistance to heat, water and many other chemicals, as well as possessing surface hardness. Alternative name: triaminotriazine. [7/9/b]

■ **melting point** Temperature at which a solid begins to liquefy, a fixed (and therefore characteristic) temperature for a pure substance.

Mendeleev's law Properties of elements listed in order of increasing relative atomic masses are generally similar every eighth element. The law was named after the Russian chemist Dimitri Mendeleev (1834–1907). Alternative name: periodic law.

mendelevium Md Radioactive element in Group IIIB of the Periodic Table (one of the **actinides**); it has several **isotopes**, with half-lives of up to 54 days. At. no. 101; r.a.m. 258 (most stable isotope).

mercaptan Alternative name for a **thiol**.

mercuric Alternative name for mercury(II) in compounds of mercury.

mercurous Alternative name for mercury(I) in compounds of mercury.

■ **mercury** Hg Dense liquid metallic element in Group IIB of the Periodic Table (a **transition element**), used in lamps, batteries, switches and scientific instruments. It alloys with most metals to form **amalgams**. Its compounds are used in drugs, explosives and pigments. At. no. 80; r.a.m. 200.59.

mercury cell *1*. **Electrolytic cell** that has a **cathode** made of mercury. *See* **polarography**. *2*. Dry cell that has a mercury electrode. Alternative name: Mallory cell.

mercury(I) chloride HgCl White crystalline compound, used as an insecticide and in a **mercury cell**. Alternative names: mercurous chloride, calomel.

mercury(II) chloride $HgCl_2$ Extremely poisonous white compound. Alternative names: mercuric chloride, corrosive sublimate.

mercury(II) oxide HgO Red or yellow compound, slightly soluble in water, which reduces to metallic mercury on heating. Alternative name: mercuric oxide.

mercury(II) sulphide HgS Red compound, which occurs naturally as cinnabar, used as a pigment (vermilion) and source of mercury.

meso- Prefix meaning middle.

mesomerism Phenomenon in which a chemical compound can adopt two or more different structures (called canonical forms) by the alteration of (covalent) bonds while the atoms in the molecules remain in the same relationship to each other (*e.g.* the Kekulé forms of **benzene**). Alternative name: resonance. *See also* **tautomerism**.

meson Member of a group of unstable **elementary particles** with masses intermediate between those of **electrons** and **nucleons**, and with positive, negative or zero charge. Mesons are emitted by **nuclei** that have been bombarded by high-energy electrons.

meta- *1*. Relating to the 1–3 carbon atoms in a **benzene ring**, abbreviated to *m*-; *e.g.* *m*-xylene is 1,3-dimethylbenzene). *2*. Relating to compounds formed by **dehydration**; *e.g.* metaphosphoric acid, made by strongly heating orthophosphoric acid. *3*. Alternative name for **metaldehyde**. *See also* **ortho-**; **para-**.

■ **metal** Any of a group of **elements** and their alloys with general properties of strength, hardness and the ability to conduct heat and electricity (because of the presence of free electrons). Most have high melting points and can be polished to a shiny finish. Metallic elements (about 80 per cent of the total) tend to form **cations**. *See also* **metalloid**. [6/7/a]

metaldehyde $(CH_3CHO)_4$. White crystalline solid, polymer of **acetaldehyde** (ethanal), used to kill slugs and as a fuel for camping and emergency stoves. Alternative names: meta, ethanal tetramer.

■ **metallic bond** Type of interatomic bonding that exists in metals.

■ **metallic crystal** Crystalline structure that is held together by **metallic bonds**.

metallocene Member of a group of chemical compounds formed between a metal and an **aromatic compound** in which the oxidation state of the metal is zero; *e.g.* **ferrocene**.

metalloid Element with physical properties resembling those of metals and chemical properties typical of non-metals (*e.g.* arsenic, germanium, selenium). Many metalloids are used in **semiconductors**.

metastatic Describing **electrons** that leave an orbital shell, either entering another shell or being absorbed into the nucleus.

methanal HCHO Alternative name for **formaldehyde**.

■ **methane** CH_4 Simplest **hydrocarbon**, an **alkane** and a major constituent of **natural gas** (up to 97%) and **coal gas**. It is an end-product of the **anaerobic** decay of plants (hence its occurrence as marsh gas in swamps); it also occurs in coal mines (where it is known as fire damp). Methane can be manufactured by the catalytic **hydrogenation** of **carbon monoxide**. It is used as a fuel, as an industrial source of hydrogen and in chemical synthesis.

■ **methanoic acid** HCOOH Alternative name for **formic acid**.

■ **methanol** CH_3OH Simplest primary **alcohol**, a poisonous liquid used as a solvent and added to ethanol to make **methylated spirits**. Alternative names: methyl alcohol, wood spirit.

methionine $CH_3S(CH_2)_2CH(NH_2)COOH$ **Sulphur**-containing **amino acid**; a constituent of many **proteins**. Alternative name: 2-amino-4-methylthiobutanoic acid.

■ **methyl alcohol** Alternative name for **methanol**.

methylamine CH_3NH_2 Simplest primary **amine**, a gas smelling like ammonia, used in making herbicides.

methylaniline Alternative name for **toluidine**.

■ **methylated spirits** General solvent consisting of absolute alcohol (ethanol) that has been denatured with **methanol** and **pyridine**; it often has added purple dye. Alternative name: denatured alcohol.

methylation Chemical reaction in which a **methyl group** is added to a chemical compound.

methylbenzene Alternative name for **toluene**.

methylbutadiene Alternative name for **isoprene**.

methyl chloroform Alternative name for **1,1,1-trichloroethane**.

methyl cyanide Alternative name for **acetonitrile**.

methylene blue Blue dye used as a pH **indicator**.

methyl group $-CH_3$ radical.

methyl methacrylate Methyl **ester** of methacrylic acid, used in the preparation of the plastic polymethyl-methacrylate (Perspex) and other acrylic resins, employed to make lenses, spectacle frames, false teeth, etc.

■ **methyl orange** Orange dye used as a pH **indicator**. [6/5/b]

methylphenol $CH_3C_6H_4OH$ Derivative of **phenol** in which one of the hydrogens of the **benzene ring** has been substituted by a **methyl group**. There are three **isomers** (ortho-, meta- and para-), depending on the positions of the substituents in the ring. Alternative name: cresol.

methylpyridine Alternative name for **picoline**.

■ **methyl red** Red dye used as a pH **indicator**.

■ **methyl salicylate** Methyl ester of **salicylic acid**, used in medicine. Alternative name: oil of wintergreen.

■ **metric system** Decimal-based system of units. *See* **SI units**.

■ **mica** Member of a group of minerals consisting of silicates. Micas have low thermal conductivity and high dielectric strength, and are widely used for electrical insulation.

■ **micro-** *1.* Metric prefix meaning a millionth; $\times 10^{-6}$ (*e.g.* microgram). It is sometimes represented by the Greek letter μ (*e.g.* μg). *2.* General prefix meaning small.

■ **milk sugar** Alternative name for **lactose**.

■ **milli-** Metric prefix meaning a thousandth; $\times 10^{-3}$ (*e.g.* milligram).

■ **milligram** (mg) Thousandth of a gram.

■ **millilitre** (ml) Thousandth of a litre, equivalent to a cubic centimetre (cc).

■ **mineral** Naturally occurring, usually crystalline, inorganic substance of more or less definite chemical composition and physical properties. Mixtures of minerals form rocks. The term is sometimes extended to include fossil fuels (coal, natural gas, petroleum).

■ **mineral acid** Inorganic acid such as sulphuric, hydrochloric or nitric acid.

■ **mineral oil** Hydrocarbon oil obtained from mineral sources or petroleum (as opposed to an animal oil or vegetable oil).

■ **mineral salts** Dissolved salts that occur in soil, derived from weathered rock and decomposed plants. They contain essential nutrients for plant growth, which are in turn utilized by herbivores (and carnivores that feed on them).

■ **miscible** Describing two or more liquids that will mutually dissolve (mix) to form a single **phase**. They can be separated by **fractional distillation**.

■ **mixture** Combination of two or more substances that do not react chemically and can be separated by physical methods (*e.g.* a **solution**). [6/6/b]

■ **moderator** Material used to slow down **neutrons** in a nuclear reactor (so that they can be captured and initiate **nuclear fission**); *e.g.* water, **heavy water, graphite**. [11/9/c] [13/10/b]

■ **molality** (m) Concentration of a solution given as the number of **moles** of solute in a kilogram of solvent.

■ **molar** Describing a quantity of a substance that is proportional to its molecular weight (a **mole**). *See* **molality**; **molarity**.

■ **molar concentration** *See* **molality**; **molarity**.

molar conductivity Electrical conductivity of an electrolyte with a concentration of 1 mole of solute per litre of solution. Expressed in siemens cm^2 $mole^{-1}$.

molar heat capacity Heat required to increase the temperature of 1 mole of a substance by 1 kelvin. Expressed in joules K^{-1} mol^{-1}.

■ **molarity** (M) Concentration of a solution given as the number of **moles** of solute in a litre of solution.

■ **molar solution** Solution that contains 1 mole of solute in 1 litre of solution.

■ **molar volume** Volume occupied by 1 mole of a substance under specified conditions.

■ **mole** Unit quantity in chemistry. It is the amount of a substance in grams that corresponds to its molecular weight, or the amount that contains particles equal in number to the **Avogadro constant**. Alternative names: mol, gram-molecule. [8/9/a]

■ **molecular distillation** Distillation at extremely low pressures. Alternative name: high-vacuum distillation.

■ **molecular formula** Method of describing the composition of a **molecule** of a chemical compound, using the chemical symbols of the constituent elements with numerical suffixes that indicate the number of atoms of each element in the molecule. *E.g.* H_2O and Na_2SO_4 are the molecular formulae of water and sodium sulphate, respectively. The molecular formula gives no indication how the component atoms are arranged. *See also* **empirical formula; structural formula.**

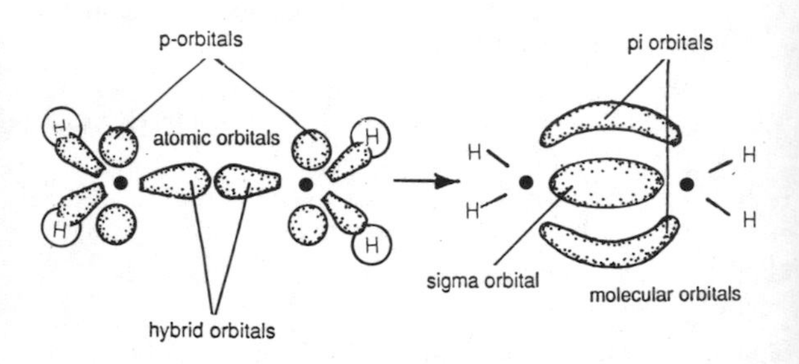

Formation of molecular orbitals

molecular orbital Region in space occupied by a pair of **electrons** that form a **covalent bond** in a **molecule**, formed by the overlap of two **atomic orbitals**.

molecular oxygen O_2 Normal **diatomic** molecular form of **oxygen**.

molecular sieve Method of separating substances by trapping (absorbing) the molecules of one within cavities of another, usually a natural or synthetic **zeolite**. Molecular sieves are used in **ion exchange, desalination** and as supports for **catalysts**.

molecule Group of atoms held together in fixed proportions by chemical bonds; the fundamental unit of a chemical compound. The simplest molecules are diatomic molecules, consisting of two atoms (*e.g.* O_2, HCl); the most complex are biochemicals and **macromolecules**. The atoms may be joined by **covalent bonds, dative bonds** or **ionic bonds**. [8/8/c]

molybdenum Mo Metallic element in Group VIB of the Periodic Table (a **transition element**), used in lamps, vacuum tubes and various alloys. At. no. 42; r.a.m. 95.94.

monatomic Describing a molecule that contains only one **atom** (*e.g.* the rare gases).

mono- Prefix meaning one (*e.g.* monobasic, monoxide).

monobasic acid Acid that on solvation produces 1 mole of **hydroxonium ion** (H_3O^+) per mole of acid; an acid with one replaceable hydrogen atom in its molecule (*e.g.* hydrochloric acid, HCl, and nitric acid, HNO_3). It cannot therefore form **acid salts**.

monocarboxylic acid Carboxylic acid with only one **carboxylic group** *e.g.* acetic (ethanoic) acid, CH_3COOH).

monoclinic Crystal form in which all three axes are unequal, with one of them perpendicular to the other two, which intersect at an angle inclined at other than a right angle.

■ **monohydrate** Chemical compound (a **hydrate**) that contains 1 mole of **water of crystallization** in each of its molecules.

■ **monohydric** Describing a chemical compound that has one **hydroxyl group** in each of its molecules (*e.g.* ethanol, C_2H_5OH, is a monohydric alcohol).

■ **monomer** Small molecule that can polymerize to form a larger molecule. *See* **polymer**.

■ **monosaccharide** $C_nH_{2n}O_n$, where $n = 5$ or 6. Member of the simplest group of **carbohydrates**, which cannot be hydrolysed to any other smaller units; e.g. the sugars glucose, fructose.

monosodium glutamate (MSG) White crystalline solid, a sodium salt of the **amino acid** glutamic acid, made from soya bean protein and used as a flavour enhancer. Eating it can cause an allergic reaction in certain susceptible people.

■ **monovalent** Having a **valence** of one. Alternative name: univalent.

montmorillonite Alternative name for **fuller's earth**.

■ **mordant** Inorganic compound used to fix colours on cloth where the cloth cannot be dyed directly.

■ **morphine** Sedative narcotic **alkaloid** drug isolated from opium, used for pain relief. Alternative name: morphia. [3/2/b] [3/6/e]

morpholine C_4H_9O **Heterocyclic** secondary **amine**, used as a solvent.

■ **multiple bond Chemical bond** that contains more electrons than a single bond (which contains 2 electrons); *e.g.* a double bond (4 electrons) or a triple bond (6 electrons).

■ **multiple proportions, law of** If two elements A and B can combine to form more than one compound, then the different masses of A that combine with a fixed mass of B are in a simple ratio. *E.g.* in carbon monoxide, CO, 16 atomic mass units of oxygen are combined with 12 units of carbon; in carbon dioxide, CO_2, 32 units of oxygen are combined with 12 units of carbon. The ratio of the oxygen masses in the two compounds is 32:16 or 2:1.

multiplication constant In a nuclear reactor, the ratio of the total number of **neutrons** produced by fission in a given time to the number absorbed or escaping in the same period. Alternative name: multiplication factor.

mustard gas $(CH_2ClCH_2)_2S$ Poisonous blistering gas used as a chemical warfare agent. Alternative name: 2,2′-dichlorodiethyl sulphide.

mutarotation Change in the **optical activity** of a solution containing photo-active substances, such as sugars.

N

nacre Mother of pearl, the inner layer of the shell of a mollusc. It is an iridescent substance, composed mainly of **calcium carbonate**.

■ **nano-** Prefix meaning a thousand-millionth; $\times 10^{-9}$ (*e.g.* nanometre)

■ **nanometre** Thousand-millionth of a metre; 10^{-9} m. It is the usual unit for wavelengths of light and interatomic bond lengths in chemistry.

■ **naphtha** Variable mixture of **hydrocarbons**, boiling range approximately 70° – 160°C, obtained from coal and petroleum and used as an industrial solvent and a raw material (feedstock) for the chemical industry. Alternative name: solvent naphtha.

naphthalene $C_{10}H_8$ Solid aromatic **hydrocarbon** that consists of two fused **benzene rings**, insoluble in water but soluble in hot ethanol. It is a starting material in the manufacture of dyes.

natron Naturally occurring hydrated **sodium carbonate**, found on the bed of dried-out soda-lakes.

■ **natural abundance** Relative proportion of one **isotope** to the total of the various isotopes in a naturally occurring sample of an element.

■ **natural gas** Mixture of hydrocarbon fuel gases, rich in **methane**, obtained from deposits that occur naturally underground. It sometimes contains **helium**.

neodymium Nd Metallic element in Group IIIB of the

Periodic Table (one of the **lanthanides**), used in special glass for lasers. At. no. 60; r.a.m. 144.24.

neon Ne Gaseous nonmetallic element in Group 0 of the Periodic Table (the **rare gases**) which occurs in trace quantities in air (from which it is extracted). It is used in discharge tubes (for advertising signs) and indicator lamps. At. no. 10; r.a.m. 20.179. [6/9/a]

neon tube Gas-discharge tube that contains **neon** at low pressure, the colour of the glow being red.

Neoprene Commercial name for the synthetic rubber polychloroprene, produced by the **polymerization** of chlorobutadiene.

neptunium Np Radioactive element in group IIIB of the Periodic Table (one of the **actinides**); it has several **isotopes**. At. no. 93; r.a.m. 237 (most stable isotope).

Nernst heat theorem If a chemical change occurs between pure crystalline solids at a temperature of absolute zero, the entropy of the final substance equals that of the initial substances.

nerve gas Chemical warfare gas that acts on the nervous system, *e.g.* by inhibition or destruction of chemicals (neurotransmitters) that convey nerve impulses across synapses between nerves.

neutral *1*. Describing a solution with **pH** equal to 7 (*i.e.* neither acidic nor alkaline). *2*. Describing a subatomic particle, atom or molecule with no residual electric charge.

neutralization Chemical reaction between an **acid** and a **base** in which both are used up; the products of the reaction are a **salt** and water; *e.g.* the reaction between hydrochloric acid (HCl) and sodium hydroxide (NaOH) to form sodium chloride (NaCl) and water (H_2O). The completion of the reaction (end-point) can be detected by an **indicator**. [7/6/a]

■ **neutral oxide** Oxide that is neither an **acidic oxide** nor a **basic oxide**; *e.g.* dinitrogen oxide (N_2O), water (H_2O). [7/5/c]

neutrino Uncharged subatomic particle with zero rest mass, a type of **lepton**, that interacts very weakly with other particles.

■ **neutron** Uncharged particle that is a constituent of the atomic **nucleus**, having a rest mass of 1.67492×10^{-27} kg (similar to that of a **proton**). Free neutrons are unstable and disintegrate by **beta decay** to a proton and an **electron**; outside the nucleus they have a mean life of about 12 minutes. [8/8/a]

neutron diffraction Technique for determining the crystal structure of solids by diffraction of a beam of **neutrons**. Similar in principle to electron diffraction, it can be used as a substitute for **X-ray crystallography**.

neutron excess Alternative name for **isotopic number**.

neutron number Number of neutrons in an atomic nucleus, the difference between the **nucleon number** of an element and its **atomic number**.

Newlands' law Alternative name for the law of **octaves**.

niacin Vitamin B_3, the only one of the B vitamins that is synthesized by animal tissues. Deficiency causes the disease pellagra. Alternative name: nicotinic acid.

■ **nickel** Ni Siver-yellow metallic element in Group VIII of the Periodic Table (a **transition element**). It is used in thermionic valves, electroplating and as a **catalyst**, and its alloys (*e.g.* stainless steel, cupronickel, German silver, nickel silver) are used in making cutlery, hollow-ware and coinage. At. no. 28; r.a.m. 58.71.

nickel carbonyl $Ni(CO)_4$ **Co-ordination compound** in which the oxidation state of nickel is zero, used as a starting material in the synthesis of a wide variety of nickel compounds.

nickel-iron accumulator Rechargable **electrolytic cell** (battery) with a positive electrode of nickel oxide and a negative electrode of iron, in a **potassium hydroxide** electrolyte. Alternative names: Edison accumulator, NiFe cell.

nickel plating Process in which a thin coating of nickel is electroplated on another metal (*see* **electroplating**). [11/7/b]

nickel silver Alloy of nickel and copper, used for coinage and for making cutlery and hollow-ware (which is often silver plated; the abbreviation EPNS stands for electroplated nickel silver). Alternative name: German silver.

nicotine Poisonous **alkaloid**, found in tobacco, which potentially binds to the receptor for the neurotransmitter acetylcholine. It is used as an insecticide.

nicotinic acid Alternative name for **niacin**.

NiFe cell Alternative name for **nickel-iron accumulator**.

niobium Nb Metallic element in Group VB of the Periodic Table (a **transition element**). Its alloys are used in high-temperature applications and superconductors. At. no. 41; r.a.m. 92.906.

nitrate Salt of **nitric acid**, containing the NO_3^- anion. Nitrates are commonly used as fertilizers, but their misuse can give rise to pollution of water supplies.

nitration Chemical reaction in which a nitro group ($-NO_2$) is incorporated into a chemical structure, to make a **nitro compound**; *e.g.* nitration of benzene (C_6H_6) gives nitrobenzene ($C_6H_5NO_2$). Nitration of organic compounds is usually achieved using a mixture of concentrated nitric and sulphuric acids (known as nitrating mixture).

nitre Old term for potassium nitrate, KNO_3, also commonly known as saltpetre.

■ **nitric acid** HNO_3 Strong extremely corrosive **mineral acid**. It is manufactured commercially by the catalytic oxidation of **ammonia** to **nitrogen monoxide** (nitric oxide) and dissolving the latter in water. Its salts are **nitrates**. The main use of the acid is in making explosives and fertilizers.

nitric oxide Alternative name for **nitrogen monoxide**.

nitrification Conversion of **ammonia** and **nitrites** to **nitrates** by the action of nitrifying bacteria. It is one of the important parts of the **nitrogen cycle**, because nitrogen cannot be taken up directly by plants except as nitrates.

nitrile Member of a group of organic compounds that contain the nitrile group ($-CN$). Alternative name: cyanide.

■ **nitrite** Salt of **nitrous acid**, containing the NO_2^- anion. Nitrites are used to preserve meat and meat products.

nitrobenzene $C_6H_5NO_2$ Aromatic liquid organic compound in which one of the hydrogen atoms in **benzene** has been replaced by a nitro group ($-NO_2$). It is used to make aniline and dyes.

nitrocellulose Alternative name for **cellulose trinitrate**.

nitro compound Organic compound in which a nitro group ($-NO_2$) is present in the basic molecular structure. Usually made by **nitration**, some nitro compounds are commercial explosives.

■ **nitrogen** N Gaseous nonmetallic element in Group VA of the

Periodic Table. It makes up about 80% of air by volume, and occurs in various minerals (particularly **nitrates**) and all living organisms. It is used as an inert filler in electrical devices and cables, and is an essential plant nutrient (*see* **fertilizer**). At. no. 7; r.a.m. 14.0067.

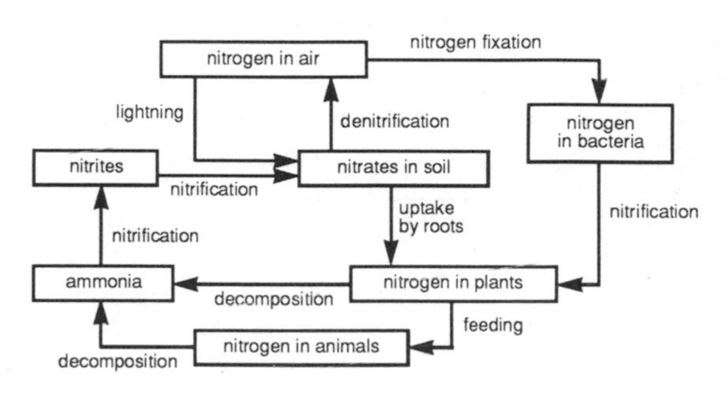

Nitrogen cycle

■ **nitrogen cycle** Circulation of nitrogen and its compounds in the environment. The main reservoirs of nitrogen are **nitrates** in the soil and the gas itself in the atmosphere (formed from nitrates by **denitrification**). Nitrates are also taken up by plants, which are eaten by animals, and after their death the nitrogen-containing **proteins** in plants and animals form **ammonia**, which **nitrification** converts back into nitrates. Some atmospheric nitrogen undergoes fixation by lightning or bacterial action, again leading to the eventual formation of nitrates (*see* **fixation of nitrogen**). [2/5/c] [2/7/a]

nitrogen dioxide NO_2 Choking brown gas made by the action of concentrated **nitric acid** on copper or of oxygen on **nitrogen monoxide** (NO). It exists in equilibrium with its dimer, dinitrogen tetroxide (N_2O_4). It is used as an **oxidizing agent** and in the manufacture of nitric acid.

■ **nitrogen fixation** *See* **fixation of nitrogen**.

nitrogen monoxide NO Colourless gas made commercially by the catalytic **oxidation** of **ammonia** and used for making **nitric acid**. It reacts with oxygen (*e.g.* in air) to form **nitrogen dioxide**. Alternative name: nitric oxide.

nitrogen oxides Compounds containing nitrogen and oxygen in various ratios, including N_2O, NO, N_2O_3, NO_2, N_2O_4, N_2O_3, N_2O_5, NO_3 and N_2O_6. The most important are dinitrogen oxide, nitrogen dioxide and nitrogen monoxide.

■ **nitroglycerine** $C_3H_5(ONO_2)_3$ Explosive oily liquid, which freezes at about 11°C. When solid, the crystals may explode at the slightest physical shock. It is used to make dynamite. Alternative names: nitroglycerin, glyceryl trinitrate, trinitroglycerine.

nitrous acid HNO_2 Weak unstable mineral acid, made by treating a solution of one of its salts (**nitrites**) with an acid.

nitrous oxide Alternative name for **dinitrogen oxide**.

NMR Abbreviation of nuclear magnetic resonance, an effect observed when radio-frequency radiation is absorbed by matter. NMR spectroscopy is used in chemistry for the study of molecular structure. It has also been introduced as a technique of diagnostic medicine.

nobelium No Radioactive element in Group IIIB of the Periodic Table (one of the **actinides**); it has various **isotopes**. At. no. 102; r.a.m. 255 (most stable isotope).

noble gas Any of the elements in group O of the Periodic Table: **helium, neon, argon, krypton, xenon** and **radon**. They have a complete set of outer electrons, which gives them great chemical stability (very few noble gas compounds are known); radon is radioactive. Alternative names: inert gas, rare gas. [6/9/a]

noble metal Highly unreactive metal, *e.g.* gold and platinum.

nodule Lump that occurs on the roots of certain plants (*e.g.* peas, beans and other legumes) which contains bacteria that are able to bring about the **fixation of nitrogen**, an important part of the **nitrogen cycle**. [2/5/c]

non-benzenoid aromatic Aromatic compound in which the number of carbon atoms in the aromatic ring is not equal to six.

non-metal Substance that does not have the properties of a **metal**. Non-metallic **elements** are usually gases (*e.g.* nitrogen, halogens, noble gases) or low-melting-point solids (*e.g.* phosphorus, sulphur, iodine). They have poor electrical and thermal conductivity, form **acidic oxides**, do not react with acids and tend to form **covalent bonds**. In ionic compounds they usually form **anions**. [6/6/c]

nonstoichiometric compound Chemical whose molecules do not contain small whole numbers of atoms. *See also* **stoichiometric compound**.

normal Describing a solution that contains 1 gram-equivalent of solute in 1 litre of solution. It is denoted by the symbol N and its multiples (thus 3N is a concentration of 3 times normal; N/10 or decinormal is a concentration of one-tenth normal).

NTP Abbreviation of normal temperature and pressure. *See also* **standard temperature and pressure** (STP).

***n*-type conductivity** Electrical conductivity caused by the flow of **electrons** in a **semiconductor**.

nuclear barrier Region of high potential energy that a charged particle must pass through in order to enter or leave an atomic nucleus.

■ **nuclear energy** Energy released during a **nuclear fission** or **nuclear fusion** process. [13/8/c]

■ **nuclear fission** Splitting of an atomic nucleus into two or more fragments of comparable size, usually as the result of the capture of a slow, or thermal, **neutron** by the nucleus. It is normally accompanied by the emission of further neutrons or **gamma-rays**, and large amounts of energy. The neutrons can continue the process as a **chain reaction**, so that it becomes the source of energy in a nuclear reactor or an atomic bomb. It may also be the 'trigger' for **nuclear fusion** in a hydrogen bomb. [8/8/b]

nuclear force Strong force that operates during interactions between certain subatomic particles. It holds together the **protons** and **neutrons** in an atomic **nucleus**.

■ **nuclear fusion** Reaction between light atomic nuclei in which a heavier nucleus is formed with the release of large amounts of energy. This process is the basis of the production of energy in stars and in the hydrogen bomb, which makes use of the fusion of **isotopes** of hydrogen to form helium. [8/10/d]

nuclear isomerism Property exhibited by nuclei with the same **mass number** and **atomic number** but different radioactive properties.

nuclear magnetic resonance (NMR) *See* **NMR**.

nuclear power Power obtained by the conversion of heat from a **nuclear reactor**, usually into electrical energy.

nuclear reaction Reaction that occurs between an atomic **nucleus** and a bombarding particle or photon, leading to the formation of a new nucleus and the possible ejection of one or more particles with the release of energy. [8/10/d]

nuclear reactor Assembly in which controlled **nuclear fission** takes place (as a **chain reaction**) with the release of heat energy. There are various types, including a **breeder reactor**, **gas-cooled reactor** and **pressurized water reactor**. [13/8/c]

nuclear transmutation Conversion of an element into another by **nuclear reaction**.

nuclear waste Radioactive by-products of a **nuclear reactor** or from the mining and extraction of nuclear fuels.

nuclear weapon Weapon whose destructive power comes from the release of energy accompanying **nuclear fission** or **fusion**; *e.g.* an atomic bomb or hydrogen bomb.

nucleon Comparatively massive particle in an atomic **nucleus**; a **proton** or **neutron**. [8/8/d]

nucleon number Total number of **neutrons** and **protons** in an atomic nucleus. *See* **mass number**. [8/8/d]

nucleophile Electron-rich chemical reactant that is attracted by **electron-deficient compounds**. Examples include an anion such as chloride (Cl^-) or a compound with a **lone pair of electrons** such as ammonia (NH_3). *See also* **electrophile**.

nucleophilic addition Chemical reaction in which a **nucleophile** adds onto an **electrophile**.

nucleophilic reagent Chemical reactant that contains electron-rich groups of atoms.

nucleophilic substitution Substitution reaction that involves a **nucleophile**.

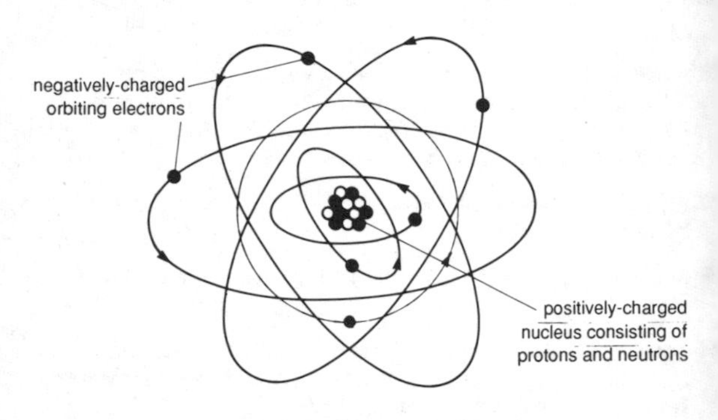

Nucleus is the centre of an atom

■ **nucleus** The most massive, central part of the atom of an element, having a positive charge given by Ze, where Z is the **atomic number** of the element and e the charge on an **electron**. It is composed of chiefly **protons** and (except for hydrogen) **neutrons**, and is surrounded by orbiting **electrons**. *See also* **isotope**. [8/8/a]

nylon Polyamide plastic made from **adipic acid** (hexanedioic acid), used mainly in making textile fibres.

O

ochre Mineral of clay and **iron(III) oxide** (Fe_2O_3), used as a light yellow to brown pigment.

octa-/octo- Prefix meaning eight.

octadecanoic acid Alternative name for **stearic acid**.

octahedral compound Chemical compound whose molecules have a central atom joined to six atoms or groups located at the vertices of an octahedron.

octahydrate Chemical containing eight molecules of **water of crystallization**.

■ **octane** C_8H_{18} Colourless liquid hydrocarbon, the eighth member of the methane series (**alkanes**). It occurs in petroleum, and is used in petrol. It has several **isomers**.

■ **octane number** Measure of **knock**-resistance in a motor fuel; the higher the octane number, the less likely it is to cause knock (preignition). Alternative name: octane rating.

octanoic acid $C_7H_{15}COOH$ Colourless oily **carboxylic acid**, used in the manufacture of dyes and perfumes. Alternative name: caprylic acid.

octaves, law of When the elements are arranged in order of their relative atomic masses, any one element has properties similar to those of the element eight places in front of it and eight places behind it in the list. It is a rejected idea of the relationship of elements. Alternative name: Newlands' law. *See also* **Periodic Table**.

■ **octet** Stable group of eight **electrons**; the outer electron configuration of most **rare gases**, and the arrangement achieved by the atoms of other elements as a result of most cases of chemical combination between them. Alternative name: electron octet.

odd-even nucleus Atomic **nucleus** with an odd number of **protons** and an even number of **neutrons**.

odd-odd nucleus Atomic **nucleus** with an odd number of both **protons** and **neutrons**.

■ **oil** *See* **fats and oils; petroleum.**

■ **oil, synthetic Hydrocarbon oil** manufactured from **coal, lignite** or **natural gas**, often by **hydrogenation**.

oil of vitriol Old name for **sulphuric acid**.

oil of wintergreen Alternative name for **methyl salicylate**.

■ **oil shale** Sedimentary rock that has a relatively high content of a bituminous substance and 30–60% organic matter. Heating it in the absence of air yields an oily substance high in sulphur and nitrogen compounds. [9/3/d]

■ **-ol** Chemical suffix that denotes an **alcohol** or **phenol**.

oleate Ester or **salt** of **oleic acid**.

■ **olefin** Alternative name for **alkene**.

oleic acid $CH_3(CH_2)_7CH=CH(CH_2)_7COOH$ **Unsaturated fatty acid** that occurs in many **fats and oils**. It is a colourless liquid that turns yellow on exposure to air.

■ **oleum** $H_2S_2O_7$ Oily solution of **sulphur trioxide** (SO_3) in concentrated **sulphuric acid**. Alternative names: disulphuric acid, fuming sulphuric acid.

oligo- Prefix that denotes small or few in number.

oligomer Polymer formed from the combination of a few **monomer** molecules.

olivine $(Mg,Fe)_2SiO_4$ Mineral which in a certain green form is classed as a gem (peridot).

omega-minus Negatively charged **elementary particle**, the heaviest **hyperon**.

■ **onium ion** Ion formed by addition of **proton** (H^+) to a molecule; *e.g.* ammonium ion (NH_4^+) from ammonia, hydroxonium ion (H_3O^+) from water.

oölite Sedimentary rock consisting of small spherical masses of calcium carbonate (oöliths).

■ **opal** Gemstone composed of non-crystalline **silica** combined with varying amounts of water. The characteristic internal play of colours in reds, blues and greens is caused by reflection of light from different layers within the stone.

■ **open-hearth process** Process for the production of **steel** from **pig iron**. Alternative name: Siemens-Martin process. [7/7/c]

optical activity Phenomenon exhibited by some chemical compounds which, when placed in the path of a beam of plane-polarized light, are capable of rotating the plane of polarization to the left (**laevorotatory**) or right (**dextrorotatory**). Alternative name: optical rotation.

optical isomerism Property of chemical compounds with the same molecular structure, but different **configurations**. Because of their molecular asymmetry they are **optically active**.

optical rotation Alternative name for **optical activity**.

orange oxide Alternative name for **uranium(VI) oxide**.

■ **orbit** Path of motion of an **electron** round the **nucleus** of an **atom**. [8/8/c]

orbital Region around the **nucleus** of an **atom** in which there is high probability of finding an **electron**. *See* **atomic orbital**; **molecular orbital**.

■ **orbital electron Electron** that **orbits** the **nucleus** of an **atom**. Alternative name: planetary electron. [8/8/c]

order of reaction Classification of chemical reactions based on the power to which the concentration of a component of the reaction is raised in the **rate law**. The overall order is the sum of the powers of the concentrations.

■ **ore** Rock or mineral from which a desired substance (usually a metal) can be extracted economically.

organic acid Organic compound that can give up **protons** to a **base**; *e.g.* a **carboxylic acid, phenol**.

organic base Organic compound that can donate a pair of **electrons** to a bond; *e.g.* an **amine**.

■ **organic chemistry** Study of **organic compounds**.

■ **organic compound** Compound of **carbon**, with the exception of its oxides and metallic carbonates, hydrogencarbonates and carbides. Other elements are involved in organic compounds, principally **hydrogen** and **oxygen** but also **nitrogen**, the **halogens**, **sulphur** and **nitrogen**.

■ **organometallic compound** Chemical compound in which a **metal** is directly bound to **carbon** in an organic group.

organosilicon compound Chemical compound in which **silicon** is directly bound to **carbon** in an organic group.

ornithine $NH_2(CH_2)_3CH(NH_2)COOH$ **Amino acid**, involved in the formation of urea in animals. Alternative name: 1,6-diaminovaleric acid.

ortho- Prefix that denotes a **benzene** compound with substituents in the 1,2 positions, abbreviated to *o-* (*e.g. o-*xylene is 1,2-dimethylbenzene). *See also* **meta-**; **para-**.

orthoarsenic acid *See* **arsenic acid**.

orthophosphoric acid Alternative name for **phosphoric(V) acid**.

osmiridium Naturally occurring alloy of **osmium** and **iridium** that is often used to make the tips of pen nibs.

osmium Os Metallic element in Group VIII of the Periodic Table (a **transition element**). The densest element, it is used in hard alloys and as a catalyst. At. no. 76; r.a.m. 190.2.

osmium(IV) oxide OsO_4 Volatile crystalline solid with a characteristic penetrating odour. Its aqueous solutions are used as a catalyst in organic reactions. Alternative name: osmium tetroxide.

■ **osmosis** Movement of a solvent from a dilute to a more concentrated solution across a **semipermeable** (or differentially permeable) **membrane**. [3/10/a]

■ **osmotic pressure** Pressure required to stop **osmosis** between a solution and pure water.

oxalate **Ester** or **salt** of **oxalic acid**.

oxalic acid $(COOH)_2.2H_2O$ White crystalline poisonous **dicarboxylic acid**. It occurs in rhubarb, wood sorrel and other plants of the oxalis group. It is used in dyeing and **volumetric analysis**. Alternative name: ethanedioic acid.

oxatyl Alternative name for **carbonyl group**.

oxidase Collective name for a group of **enzymes** that promote **oxidation** within plant and animal cells.

■ **oxidation** Process that involves the loss of **electrons** by a substance; the combination of a substance with **oxygen**. It may occur rapidly (as in combustion) or slowly (as in rusting and other forms of corrosion). [7/7/b]

■ **oxidation number** Number of **electrons** that must be added to a **cation** or removed from an **anion** to produce a neutral **atom**. An oxidation number of zero is given to the **elements** themselves. In compounds, a positive oxidation number indicates that an element is in an oxidized state; the higher the oxidation number, the greater is the extent of oxidation. Conversely, a negative oxidation number shows that an element is in a reduced state.

■ **oxidation-reduction reaction** Alternative name for a **redox reaction**. [7/7/b]

■ **oxide** Compound of **oxygen** and another element, usually made by direct combination or by heating a **carbonate** or **hydroxide**. *See also* **acidic oxide**; **basic oxide**; **neutral oxide**. [7/5/c]

■ **oxidizing agent** Substance that causes **oxidation**. Alternative name: electron acceptor.

oxime Compound containing the oximino group $=NOH$, derived by the condensation of an **aldehyde** or **ketone** with hydroxylamine (NH_2OH); *e.g.* acetaldehyde (ethanal), CH_3CHO, forms $CH_3CH=NOH$.

oxo process Reaction that involves the addition of **hydrogen** and the formyl group $-CHO$, derived from hydrogen and **carbon monoxide**, to an **alkene** in the presence of a **catalyst** (usually cobalt). It is used in the conversion of alkenes into **aldehydes** and **alcohols**. Alternative name: oxo reaction.

2-oxopropanoic acid Alternative name for **pyruvic acid.**

■ **oxy-acetylene burner** Device for obtaining a very high-temperature flame (3,480°C) by the combination of **oxygen** and **acetylene** in the correct proportions. It is used in cutting and welding of metals.

oxyacid Acid in which the acidic (*i.e.* replaceable) hydrogen atom is part of a **hydroxyl group** (*i.e.* organic **carboxylic acids** and **phenols**, and inorganic acids such as **phosphoric(V) acid** and **sulphuric acid**).

■ **oxygen** O Gaseous nonmetallic element in Group VIA of the Periodic Table. A colourless odourless gas, it makes up about 20% of air by volume, from which it is extracted, and is essential for life. It is the most abundant element in the Earth's crust, occurring in all water and most rocks. It is used in welding, steel-making and as a rocket fuel. It has a triatomic **allotrope, ozone** (O_3). At. no. 8; r.a.m. 15.9994.

■ **ozone** O_3 Allotrope of **oxygen** that contains three atoms in its molecule. It is formed from oxygen in the upper atmosphere by the action of ultraviolet light, where it also acts as a shield that prevents excess ultraviolet light reaching the Earth's surface. It is a powerful **oxidizing agent**, often used in organic chemistry.

■ **ozone layer** Layer in the upper atmosphere at a height of between 15 and 30 km, where **ozone** is found in its greatest concentration. It filters out ultraviolet radiation from the Sun which would otherwise be harmful. If this layer were destroyed, or depleted to a great extent, life on Earth would be endangered. Alternative name: ozonosphere. [5/9/a]

P

packing fraction Difference between the actual mass of an **isotope** and the nearest whole number divided by the **mass number**.

■ **paint** Suspension of powdered colouring matter (**pigment**) in a liquid, used for decorative and protective coatings of surfaces. In an oil paint, the liquid contains solvents and a drying oil or synthetic resin. In emulsion paint, the liquid is mainly water containing a dispersion of resin.

palladium Pd Silver-white metallic element in Group VIII of the Periodic Table (a **transition element**), used as a **catalyst** and in making jewellery. At. no. 46; r.a.m. 106.4.

palmitic acid $C_{15}H_{31}COOH$ Long-chain **carboxylic acid** which occurs in oils and fats (*e.g.* palm oil) as its glyceryl **ester**, used in making **soap**.

■ **papain** Proteolytic **enzyme**, which digests proteins, found in various fruits and used as a meat tenderizer.

■ **paper chromatography** Type of **chromatography** in which the mobile phase is liquid and the stationary phase is porous paper. Compounds are separated on the paper, and can then be identified.

para- *1.* Prefix that denotes the form of a diatomic molecule in which both nuclei have opposite spin directions.
2. Referring to the 1,4 positions in the benzene ring (can be abbreviated to *p*-); *e.g.* *p*-xylene is 1,4-dimethylbenzene. *See also* **meta-**, **ortho-**.

■ **paraffin** **Alkane**, a member of series of **hydrocarbons** which include gases as well as liquids and solids, obtained by the

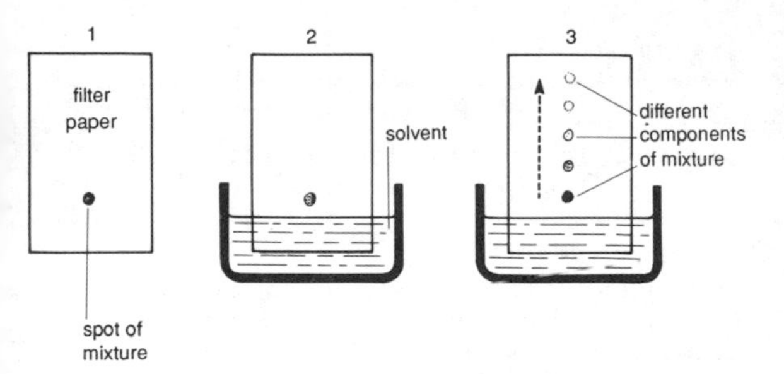

Simple paper chromatography

distillation of **petroleum**. Paraffins are used as solvents and fuels.

■ **paraffin oil** Alternative name for **kerosene**.

■ **paraffin wax** Mixture of solid **paraffins** (alkanes) which takes the form of a white translucent solid that melts below 80°C. It is used to make candles. Alternative name: petroleum wax.

paraformaldehyde Alternative name for **polymethanal**.

paraldehyde $(CH_3CHO)_3$ Cyclic **trimer** formed by the polymerization of **acetaldehyde** (ethanal), used as a sleep-inducing drug. Alternative name: ethanal trimer. *See also* **metaldehyde**.

partial pressure *See* **Dalton's law of partial pressure.**

■ **particle** Minute portion of matter, often taken to mean an **atom, molecule** or **elementary particle** or **subatomic particle.** [8/4/a]

partition coefficient Ratio of the concentrations of a single **solute** in two immiscible **solvents**, at equilibrium. It is independent of the actual concentrations.

passivity Corrosion-resistance (and unreactivity) of a metal that results from a thin surface layer of oxide. *E.g.* aluminium is passive because of such a layer, which can be thickened by **anodizing.**

■ **pasteurization** Process of heating food or other substances under controlled conditions. It was developed by the French chemist Louis Pasteur (1822–95) to destroy germs. It is widely used in industry, *e.g.* production of milk and wine.

Pauli exclusion principle No two **electrons** can be assigned the same set of **quantum numbers**; hence there can be only two electrons in any one atomic **orbital**. It was named after the Austrian-born American physicist Wolfgang Pauli (1900–58). Alternative name: exclusion principle.

***p*-block elements** 30 nonmetallic elements that form Groups IIB, IVB, VB, VIB, VIIB and 0 of the Periodic Table (helium is usually excluded), so called because their 1 to 6 outer electrons occupy ***p*-orbitals.**

■ **pearl** Lustrous, often spherical, accretion formed from the layering of **nacre** (calcium carbonate) on a foreign particle inside the shells of certain molluscs, *e.g.* oysters.

pearl spar Alternative name for **dolomite.**

■ **peat** Brown or black fibrous material formed, in surface deposits, by the partial decomposition of plant remains. It is used in making manures and composts, in making charcoal, and as a heat-insulating material. Dried peat can be used as a fuel.

■ **pectin** Complex **polysaccharide** derivative present in plant cell walls, to which it gives rigidity. It can be converted to a gel form in sugary acid solution.

■ **pentahydrate** Chemical containing five molecules of **water of crystallization**; *e.g.* $CuSO_4.5H_2O$.

pentane C_5H_{12} Liquid **alkane**, extracted from petroleum and used as a solvent and in organic synthesis. It has three **isomers**.

pentanoic acid $CH_3(CH_2)_3COOH$ Liquid **carboxylic acid** with a pungent odour, used in perfumes. Alternative name: valeric acid.

■ **pentavalent** Having a **valence** of five.

pentose sugar Monosaccharide carbohydrate (sugar) that contains five carbon atoms and has the general formula $C_5H_{10}O_5$; *e.g.* **ribose** and **xylose**. Alternative name: pentaglucose.

pentyl group Alternative name for **amyl group**.

■ **pepsin Enzyme** produced in the stomach which, under acid conditions, brings about the partial **hydrolysis** of **polypeptides** (thus helping in the digestion of **proteins**).

■ **peptidase Enzyme**, often secreted in the body (*e.g.* by the intestine), which degrades **peptides** into free **amino acids**, thus completing the digestion of **proteins**.

peptide Organic compound that contains two or more **amino acid** residues joined covalently through peptide bonds (–NH–CO–) by a condensation reaction between the carboxyl group of one amino acid and the amino group of another. Peptides polymerize to form **proteins**.

■ **percentage composition** Make-up of a chemical compound expressed in terms of the percentages (by mass) of each of its component elements. *E.g.* ethane (C_2H_6), ethene (C_2H_4) and ethyne (C_2H_2) all consist of carbon and hydrogen. Their approximate percentage compositions are:

ethane: 80% carbon, 20% hydrogen
ethene: 85% carbon, 15% hydrogen
ethyne: 92% carbon, 8% hydrogen

perdisulphuric(VI) acid H_2SO_5 White crystalline compound, used as a powerful **oxidizing agent**. Alternative names: Caro's acid, persulphuric acid, peroxomonosulphuric acid.

perfect gas Alternative name for **ideal gas**.

■ **period** One of the seven horizontal rows of elements in the **Periodic Table**.

■ **periodicity** Regular increases and decreases of physical values for elements known to have similar chemical properties.

■ **periodic law** Properties of elements are a periodic function of their atomic numbers. Alternative name: Mendeleev's law.

■ **Periodic Table** Arrangement of elements in order of increasing **atomic number**, with elements having similar properties, *i.e.* in the same family, in the same vertical column (group). Horizontal rows of elements are termed periods. [6/7/c]

permanent gas Gas that is incapable of being liquefied by **pressure** alone; a gas above its critical temperature.

■ **permanent hardness (of water)** Hardness that is not destroyed by boiling the water. It is caused by the presence of calcium or magnesium salts. *See* **hardness of water**. [5/6/a]

Permutit Trade name for a type of **zeolite** used in water softeners. [5/6/a]

peroxide *1.* **Oxide** of an element containing more **oxygen** than does its normal oxide. *2.* Oxide, containing the O_2^{2-} ion, that yields **hydrogen peroxide** on treatment with an **acid**. Peroxides are powerful **oxidizing agents**.

peroxomonosulphuric acid Alternative name for **perdisulphuric(VI) acid**.

Perspex Trade name for the plastic **polymethyl methacrylate**.

persulphate salt of **perdisulphuric(VI) acid**, used as an oxidizing agent.

persulphuric acid Alternative name for **perdisulphuric(VI) acid**.

pesticide Compound used in agriculture to destroy organisms that can damage crops or stored food, especially insects and rodents. Pesticides include fungicides, herbicides and insecticides. The effects of some of them, *e.g.* organic chlorine compounds such as DDT, can be detrimental to the ecosystem. [5/10/a]

petrochemical Any of a range of chemicals derived from **petroleum** or **natural gas**.

petrol Liquid fuel (for internal combustion engines) consisting of a mixture of **alkanes** in the range pentane (C_5H_{12}) to decane ($C_{10}H_{22}$), made by distillation and **cracking** of **petroleum**. Alternative names: gasoline, motor spirit. *See also* **diesel fuel**. [7/9/a]

petroleum Mineral oil, a mixture of **hydrocarbons**, which is usually greenish, brown or black in the crude state. It is the source of various fuels and an important natural raw material in the chemical industry.

■ **petroleum wax** Alternative name for **paraffin wax**.

■ **pewter Alloy** of **tin** (80–90%) and **lead** (10–20%), used to make tableware and jewellery.

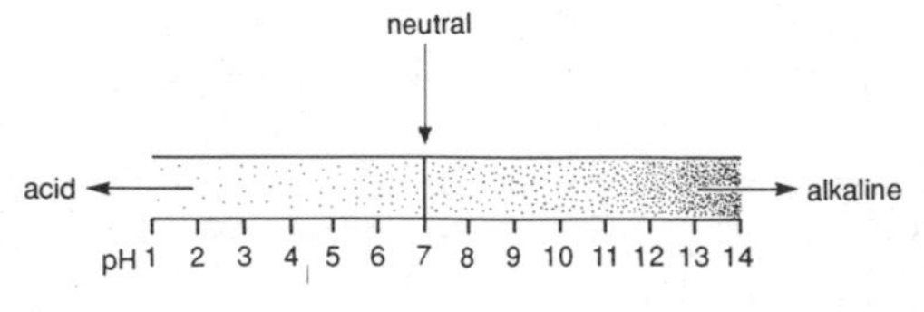

pH scale is neutral at pH 7

■ **pH** Hydrogen ion concentration (grams of hydrogen ions per litre) expressed as its negative logarithm; a measure of acidity and alkalinity. For example, a hydrogen ion concentration of 10^{-3} grams per litre corresponds to a pH of 3, and is acidic. A pH of 7 is neutral; a pH of more than 7 is alkaline. [6/5/b]

pharmacology Study of the properties, manufacture and reactions of drugs.

phase Any homogeneous and physically distinct part of a chemical system that is separated from other parts of the system by definite boundaries, *e.g.* ice mixed with water. Solids, liquids and gases make up different phases.

phase rule The number of **degrees of freedom** (F) of a **heterogeneous** system is related to the number of components (C) and of **phases** (P) present at equilibrium by the equation $P + F = C + 2$.

phenol C_6H_5OH Colourless crystalline solid which turns pink on exposure to air and light. It has a characteristic, rather sweet odour. It is used as an antiseptic and disinfectant, and in the preparation of dyes, drugs, etc. Alternative names: carbolic acid, hydroxybenzene. Other compounds with one or more **hydroxyl groups** bound directly to a **benzene ring** are also known as phenols. They give reactions typical of **alcohols** (*e.g.* they form **esters** and **ethers**), but they are more acidic and form salts by the action of strong alkalis.

■ **phenolphthalein** Organic compound that is used as a laxative and as an **indicator** in **volumetric analysis**. It is red in alkalis and colourless in acids.

phenylalanine $C_6H_5CH_2CH(NH_2)COOH$ Essential amino acid that possesses a **benzene ring**.

phenylamine Alternative name for **aniline**.

phenylethylene Alternative name for **styrene**.

phenyl group C_6H_5- **Monovalent radical** derived from **benzene**.

phenyl methyl ketone Alternative name for **acetophenone**.

3-phenylpropenoic acid Alternative name for cinnamic acid.

phosgene $COCl_2$ Colourless gas with penetrating and suffocating smell. It has been employed as a war gas, and is now used in organic synthesis. Alternative name: carbonyl chloride.

■ **phosphate** Salt of **phosphoric(V) acid**, containing the ion PO_4^{3-}. Some phosphates are of enormous commercial and practical importance, *e.g.* ammonium phosphate **fertilizers**, and alkali phosphate **buffers**. Alternative name: orthophosphate. Because phosphoric(V) acid is a tribasic acid, it also forms hydrogenphosphates, HPO_4^{2-}, and dihydrogenphosphates, $H_2PO_4^-$. [2/5/c]

phosphide Chemical compound of **phosphorus** and another element.

phosphine PH_3 Colourless poisonous gas, often spontaneously inflammable because of impurities. Alternative name: phosphorus trihydride, phosphorus(III) hydride.

phospholipid Member of a class of complex **lipids** that are major components of cell membranes. They consist of molecules containing a phosphoric(V) acid **ester** of **glycerol**, the remaining **hydroxyl groups** of the glycerol being esterified by **fatty acids**. Alternative names: phosphoglyceride, phosphatide, glycerol phosphatide.

phosphonium ion Ion PH_4^+.

phosphor Substance capable of luminescence or phosphorescence, as used to coat the inside of a television screen or fluorescent lamp.

phosphoric(V) acid H_3PO_4 Tribasic acid, a colourless crystalline solid, made by dissolving **phosphorus(V) oxide** in water. It is used to form a corrosion-resistant layer on steel. Alternative names: phosphoric acid, orthophosphoric acid.

■ **phosphorus** P Nonmetallic element in Group VA of the Periodic Table. It exists as several **allotropes**, chief of which are red phosphorus and the poisonous and spontaneously inflammable white or yellow phosphorus. It occurs in many minerals (particularly **phosphates**) and all living organisms; it is an essential nutrient for plants (*see* **fertilizer**). Phosphorus is made by heating calcium phosphate (with carbon and sand) in an electric furnace. It is used in matches and for making fertilizers. At. no. 15; r.a.m. 30.9738.

phosphorus(III) bromide PBr_3 Colourless liquid, used in organic synthesis to replace a **hydroxyl group** with a **bromine** atom. Alternative name: phosphorus tribromide.

phosphorus(V) bromide PBr_5 Yellow crystalline solid, used as a brominating agent. Alternative name: phosphorus pentabromide.

phosphorus(III) chloride PCl_3 Colourless fuming liquid, used to make organic compounds of phosphorus. Alternative name: phosphorus trichloride.

phosphorus(V) chloride PCl_5 Yellowish-white crystalline solid, used as a chlorinating agent. Alternative name: phosphorus pentachloride.

phosphorus(III) hydride Alternative name for **phosphine**.

phosphorus(III) oxide P_2O_3 White waxy solid which readily reacts with oxygen to form **phosphorus(V) oxide**. Alternative name: phosphorus trioxide.

phosphorus(V) oxide P_2O_5 **Hygroscopic** white powder which readily reacts with water to form **phosphoric(V) acid**. It is used as a **desiccant**. Alternative name: phosphorus pentoxide.

■ **photochemical reaction** Chemical reaction that is initiated by the absorption of light. The most important phenomenon of this type is photosynthesis. It is also the basis of photography. [3/6/b]

■ **photochemistry** Branch of chemistry concerned with the action of light in initiating chemical reactions.

■ **photography** Process of taking photographs by the chemical action of light or other radiation on a sensitive plate or film made of glass, celluloid or other transparent material coated with a light-sensitive emulsion. Light causes changes in particles of silver salts in the emulsion which, after development (in a **reducing agent**), form grains of dark metallic silver to produce a negative image. Unaffected silver salts are removed by fixing (in a solution of ammonium or sodium thiosulphate).

photoionization Ionization of atoms or molecules by light or other electromagnetic radiation.

photolysis Photochemical reaction that results in the decomposition of a substance.

■ **pH scale** Scale that indicates the acidity or alkalinity of a solution. *See* **pH**. [6/5/b]

phthalic acid $C_6H_4(COOH)_2$ White crystalline solid which on heating converts to its **anhydride**. It is used in organic synthesis and to make **polyester** resins. Alternative name: benzene-1,2-dicarboxylic acid.

phthalocyanine Member of an important class of synthetic organic dyes and pigments. They are blue to green and used for colouring paints, printing inks, synthetic plastics and fibres, rubber, etc.

■ **physical change** Reversible alteration in the properties of a substance that does not affect the composition of the substance itself (as opposed to a chemical change, which is difficult to reverse and in which composition is affected).

■ **physical chemistry** Branch of chemistry concerned with the physical properties of substances.

picoline $CH_3C_6H_4N$ **Heterocyclic** liquid organic **base** that exists in three isomeric forms. It occurs in coal-tar and bone oil, and is used in organic synthesis. Alternative name: methylpyridine.

picrate *1.* Salt of **picric acid**. *2.* Compound (a charge-transfer complex) formed between **picric acid** and an **aromatic compound, amine** or **phenol**; picrates are frequently used to identify these classes of compounds.

picric acid $C_6H_2(NO_3)_3OH$ Yellow crystalline solid obtained by nitrating phenol sulphonic acid. It has been used as a dye and as an explosive. Alternative name: 2,4,6-trinitrophenol.

■ **pig iron** High-carbon **iron** made by the smelting (reduction) of iron ore in a **blast furnace**.

■ **pigment** Insoluble colouring material, used for imparting various colours to **paints**, paper, **polymers**, etc. (soluble colouring materials are dyes). Some naturally occurring coloured substances are also known as pigments; *e.g.* green chlorophyll in plants and red haemoglobin in blood.

pine cone oil Alternative name for **turpentine**.

■ **pinking** Alternative name for knocking (preignition). *See* **knock**.

■ **pipette** Device for transferring a known volume of liquid. It consists of a glass tube, often with a swelling at its centre, and may have a rubber bulb or glass 'cylinder' at one end. Pipettes are used in **volumetric analysis**.

pitchblende Glossy black mineral consisting mainly of **uranium(VI) oxide**. It is the principal ore of **radium** and **uranium**. Alternative name: uranite.

pK value Negative logarithm of the **equilibrium constant** for the **dissociation** of an **electrolyte** in aqueous solution.

Planck's constant (*h*) Fundamental constant that relates the energy of a **quantum** of **radiation** to the frequency of the oscillator that emits it. The relationship is $E = h\upsilon$, where E is the energy of the quantum and υ is its frequency. Its value is $6.626196 \times 10_{-34}$ joule second. It was named after the German physicist Max Planck (1858–1947).

■ **planetary electron** Alternative name for **orbital electron**.

plasma State of matter in which the atoms or molecules of a substance are broken into **electrons** and positive **ions**. All substances pass into this state of matter when heated to a very high temperature, *e.g.* in an electric arc or in the interior of a star.

■ **plaster of Paris** $CaSO_4.\frac{1}{2}H_2O$ Calcium sulphate hemihydrate, obtained by heating **gypsum**. When water is added, it sets hard, re-forming gypsum. In doing so, it does not expand or contract much, and is therefore valuable as a moulding material, particularly as a splint for broken bones and in the building industry.

■ **plastic** Member of a large class of substances that under heat and pressure become capable of flow and can then be given a shape which is retained when the heat and pressure are removed. Plastics that re-soften on heating are termed thermoplastic; those that do not are thermosetting. These substances are derived from animal or vegetable sources, coal-tar or **petroleum**. Most plastics are **polymers**. [7/9/b]

plasticizer Compound added to **plastics** to make them soft and readily workable.

■ **platinum** Pt Valuable silver-white metallic element in Group VIII of the Periodic Table (a **transition element**). It is used for making jewellery, electrical contacts and in scientific apparatus. At. no. 78; r.a.m. 195.09.

platinum black Finely divided form of **platinum**.

plumbic Alternative name for **lead(IV)**.

plumbous Alternative name for **lead(II)**.

■ **plutonium** Pu Radioactive element in Group IIIB of the Periodic Table (one of the **actinides**), produced from uranium-238 in a **breeder reactor**. It has several **isotopes**, some of which (*e.g.* Pu-239) undergo **nuclear fission**; all are very poisonous. At. no. 94; r.a.m. 244 (most stable isotope).

Symbol on containers of poisonous chemicals

■ **poison** *1.* Substance that destroys the activity of a **catalyst**. *2.* Substance that when introduced into a living organism in any way destroys life or causes injury to health; a toxin.

polar bond **Covalent bond** in which the bonding **electrons** are not shared equally between the two **atoms**.

polar crystal **Crystal** that has **ionic bonds** between its atoms. Alternative name: ionic crystal.

polarimeter Instrument for measuring the **optical activity** of a substance. Alternative name: polariscope.

polarimetry Measurement of **optical activity**, used in chemical analysis.

■ **polarization** *1.* Separation of the positive and negative charges of a molecule. *2.* Formation of gas bubbles or a film of deposit on an electrode of an **electrolytic cell**, which tends to impede the flow of current.

polarized light Light waves (which normally oscillate in all possible planes) with fixed orientation of the electric and magnetic fields. It may be created by passing the light through a polarizer consisting of a plate of tourmaline crystal cut in a special way, by using a Polaroid sheet.

polar molecule Molecule that is polarized (*see* **polarization**) even in the absence of an electric field.

polarography Method of chemical analysis for substances in dilute solution in which current is measured as a function of potential between mercury electrodes in an **electrolytic cell** containing the solution.

■ **pollution** Harmful changes or presence of undesirable substances in the environment which result from mankind's industrial or social activities. Pollution of the atmosphere includes the presence of sulphur dioxide, which causes acid rain, and of chlorofluorocarbons (CFCs), which have been linked to the depletion of the **ozone layer** in the stratosphere. River pollution is caused mainly by agricultural run-off (*e.g.*

of fertilizers or slurry) or discharge of chemicals. Pollution of the oceans is caused by oil spillage or dumping of untreated sewage, industrial wastes or chemical wastes. [5/5/a]

polonium Po Radioactive metallic element in Group VIA of the Periodic Table, used as a source of alpha-particles. At. no. 84; r.a.m. 209 (most stable isotope).

■ **poly-** Prefix meaning many (*e.g.* polyamide, polymer).

polyamide Condensation **polymer** in which the units are linked by **amide** groups (– CONH –); *e.g.* hair, wool fibres, nylon.

polybasic Describing an **acid** with two or more acidic (replaceable) hydrogen atoms in its molecule; *e.g.* phosphorus(V) (orthophosphoric) acid, H_3PO_4, with three replaceable hydrogens, is tribasic.

■ **polycarbonate** Linear low-crystalline **thermoplastic** in which the linking elements are **carbonate** groups. Polycarbonates are used in making electrical connectors and soft-drink bottles.

polychloroethene Alternative name for **polyvinyl chloride** (PVC).

polycyclic Describing a substance that has more than one ring of atoms in its molecule.

■ **polyester** Condensation **polymer** formed from a **polyhydric alcohol** and a **polybasic acid**. Polyesters are used in the manufacture of fibres.

polyethylene Thermoplastic produced by the **polymerization** of **ethene** (ethylene). It is used for making film and sheeting for bags and wrappers and for making moulded articles. Alternative names: polyethene, polythene.

polyhydric Containing a number of **hydroxyl groups**.

polyhydric alcohol Alcohol that contains three or more **hydroxyl groups**.

■ **polymer** Long-chain molecule built up of a number of smaller molecules, called **monomers**, joined together by **polymerization**. Natural polymers include starch, cellulose and rubber. Synthetic polymers include all kinds of **plastics**. [7/9/b]

■ **polymerization** Process of joining together of small molecules, called **monomers**, to form larger molecules, **polymers** (often in the presence of a **catalyst**). In condensation polymerization, two types of monomer molecules condense to form long chains, with the elimination of a small molecule (such as water). In addition polymerization, long chains are formed by molecules of a single monomer joining together. [7/9/b]

polymethanal Polymer formed from methanal (**formaldehyde**). Alternative name: paraformaldehyde.

polymethyl methacrylate Transparent colourless **thermoplastic**. Its optical properties of high transmission of light and high internal reflection, coupled with great strength, are responsible for its use in place of **glass**. Alternative name: Perspex.

polymorphism Occurrence of a substance in more than one crystalline form.

polypeptide Chain of **amino acids** which is a basic constituent of **proteins**. It may be broken down by **enzyme** action (digestion) to form **peptides**. The linking and folding of polypeptides makes up the three-dimensional structure of a protein.

polyphenylethene Alternative name for **polystyrene**.

■ **polypropene Thermoplastic** produced by the **polymerization** of **propene** (propylene). It is used to produce moulded articles and can be made into a fibre. Alternative name: polypropylene.

polypropylene Alternative name for **polypropene**.

■ **polysaccharide** High molecular weight **carbohydrate**, linked by **glycoside** bonds, that yields a large number of **monosaccharide** molecules (*e.g.* simple sugars) on **hydrolysis** or **enzyme** action. The most common polysaccharides have the general formula $(C_6H_{10}O_5)_n$; e.g. starch, cellulose, etc.

■ **polystyrene Thermoplastic** produced by the **polymerization** of **styrene** (phenylethene). It is used to produce moulded articles and, as a foam (expanded polystyrene), for ceiling tiles, insulation and packaging. Alternative name: polyphenylethene. [7/9/b]

polytetrafluoroethene (PTFE) **Thermoplastic** produced by the **polymerization** of **tetrafluoroethene** (tetrafluoroethylene). It is inert, very stable and has anti-stick properties. It is used in non-stick coatings on cooking utensils and as an insulator. Trade names: Fluon, Teflon. Alternative name: polytetrafluoroethylene.

■ **polythene** Alternative name for **polyethylene**.

■ **polyurethane** Member of a family of **polymers** (plastics) in which the formation of the **urethane** group is an important step in **polymerization**. They are used for the manufacture of foams and coatings. Alternative name: urethane resin.

■ **polyvalent** *1.* Having a **valence** of more than one. *2.* Having more than one valence.

polyvinyl acetate (PVA) **Thermoplastic** produced by the **polymerization** of vinyl acetate (ethanoate), $CH_2=CHOOCCH_3$. It is used in adhesives and for coating paper and fabrics.

■ **polyvinyl chloride** (PVC) **Thermoplastic** produced by the **polymerization** of vinyl chloride (chloroethene), ($CH_2=CHCl$). It is used as electrical insulation for wires and cables, for making pipes and gramophone records, and for making waterproof clothing. Alternative name: polychloroethene. [7/9/b]

■ **porcelain** Ceramic made from **quartz**, white **kaolin**, **marble** and **feldspar**. The plastic paste is first moulded and then fired.

porphyrin Member of an important class of naturally occurring organic **pigments** derived from four **pyrrole** rings. Many form complexes with metal ions, as in *e.g.* chlorophyll, haem, cytochrome, etc.

positron Elementary particle which has a mass equal to that of an **electron**, and an electrical charge equal in magnitude, but opposite in sign, to that of the electron.

post-actinide Any of the elements with an atomic number greater than that of lawrencium (103), the last of the **actinides**. The post-actinides 104 (hahnium) and 105 (rutherfordium or kurchatovium) have been prepared in minute quantities by bombarding an actinide with atoms of elements such as carbon, oxygen and neon. Element 106 has also been claimed.

■ **potash** Substance that contains **potassium**, particularly **potassium carbonate**.

■ **potassium** K Highly reactive silver-white metallic element in Group IA of the Periodic Table (the **alkali metals**). Its compounds occur widely (particularly the **chloride**) and have many uses; potassium is an essential nutrient for plants (*see* **fertilizer**). The metal is used as a coolant in **nuclear reactors**. At. no. 19; r.a.m. 39.102.

potassium bicarbonate Alternative name for **potassium hydrogencarbonate**.

potassium bromide KBr White crystalline salt, used in medicine and photography.

potassium carbonate K_2CO_3 White granular solid, used in the manufacture of **glass** and **soap**. Alternative name: potash.

potassium chloride KCl Colourless or white crystalline salt, used as a **fertilizer** and as a dietary salt (sodium chloride) substitute when **sodium** intake must be limited.

potassium cyanide KCN White poisonous solid, used in metallurgy and electroplating.

potassium ferricyanide $K_3Fe(CN)_6$ Red crystals, used as a chemical reagent and in the manufacture of **pigments** (*e.g.* Prussian blue). Alternative name: potassium hexacyanoferrate(III).

potassium ferrocyanide $K_4Fe(CN)_6.3H_2O$ Yellow crystals. Alternative name: potassium hexacyanoferrate(II).

potassium hydrogencarbonate $KHCO_3$ White granular solid, used in pharmaceuticals. Alternative name: potassium bicarbonate.

potassium hydrogentartrate $HOOC(CHOH)_2$ COOK White crystalline powder, used in baking powder. Alternative name: cream of tartar.

■ **potassium hydroxide** KOH Strongly **hygroscopic** white solid. A strong **alkali**, it is used in the manufacture of soft **soaps**. Alternative name: caustic potash.

potassium iodide KI Colourless crystalline salt, used in chemical analysis and organic synthesis. Its solution dissolves iodine.

■ **potassium manganate(VII)** Alternative name for **potassium permanganate**.

■ **potassium nitrate** KNO_3 Colourless crystalline salt, a powerful **oxidizing agent**. It is used in the manufacture of **glass** and explosives, and as a food preservative. Alternative names: saltpetre, nitre.

■ **potassium permanganate** $KMnO_4$ Purple crystals, a powerful **oxidizing agent**. It is used in the manufacture of chemicals, as a disinfectant and fungicide, and in **volumetric analysis**. Alternative names: permanganate of potash, potassium manganate(VII).

potassium thiocyanate KSCN Colourless **hygroscopic** solid, used in solution to test for iron(III) (ferric) compounds, which give a blood-red colour.

■ **powder metallurgy** Science of producing metal powders and of using them for the production of shaped objects (*see* **sintering**).

praseodymium Pr Metallic element in Group IIIB of the Periodic Table (one of the **lanthanides**). At. no. 59; r.a.m. 140.907.

■ **precipitate** Solid that forms in and settles out from a solution.

■ **precipitation** Process of **precipitate** formation. *See* **double decomposition**. [7/8/b]

precursor Intermediate substance from which another is formed in a chemical reaction.

pressure (p) Force applied to, or distributed over, a surface, measured in force f per unit area a; $p = f/a$. The SI unit of pressure is the pascal; other units include bars, millibars, atmospheres and millimetres of mercury (mm Hg). [6/6/e]

pressurized water reactor (PWR) **Nuclear reactor** in which the heat generated in the nuclear core is removed by water (reactor coolant), circulating at high pressure to prevent it boiling.

primary alcohol/amine Alcohol or **amine** with only one **alkyl** or **aryl group**.

primary cell Electrolytic cell (battery) in which the chemical reactions that cause the current flow are not readily reversible and the cell cannot easily be recharged, *e.g.* a **dry cell**. *See also* **secondary cell**.

probability distribution of electrons Probability that an **electron** within an atom will be at a certain point in space at a given time. It predicts the shape of an atomic **orbital**.

producer gas Mixture of the gases **hydrogen**, **nitrogen** and **carbon monoxide** made by passing steam and air through red-hot coke. Alternative name: air gas.

product Substance formed as a result of a chemical change.

proenzyme Alternative name for **zymogen**.

proline White crystalline **amino acid** that occurs in most **proteins**.

promethium Pm Radioactive metallic element in Group IIIB of the Periodic Table (one of the **lanthanides**). It has several **isotopes** (none of which occurs naturally), with half-lives of up to 20 years. At. no. 61; r.a.m. 145 (most stable isotope).

promoter Substance used to enhance the efficiency of a **catalyst**. Alternative name: activator. [7/7/a]

■ **proof** Measure of the **ethanol** (ethyl alcohol) content of a solution (gunpowder moistened with a 100% proof spirit will just ignite). A 100% proof solution is 57.1% ethanol by volume or 49.3% alcohol by weight.

propanal CH_3CH_2CHO Liquid **aldehyde,** used in the manufacture of **plastics**. Alternative name: propionaldehyde, propyl aldehyde.

■ **propane** C_3H_8 Gaseous **alkane,** easily liquefied under pressure and used as a portable supply of fuel.

propanoic acid CH_3CH_2COOH Liquid **carboxylic acid,** whose calcium salt is used as a food additive. Alternative name: propionic acid.

propanol Alcohol that occurs as two isomers. *1. n*-propanol C_3H_7OH is a colourless liquid, used as a solvent and in making toilet preparations. Alternative names: *n*-propyl alcohol, propan-1-ol. *2.* Isopropanol $(CH_3)_2CHOH$ is also a colourless liquid, used for preparing **esters, acetone** (propanone), and as a solvent. Alternative names: isopropyl alcohol, propan-2-ol.

2-propanone Alternative name for **acetone.**

■ **propellant** *1.* Explosive used to propel bullets and shells, or give thrust to solid-fuel rockets. *2.* Gas used in an **aerosol** to expel the contents through an atomizing jet.

2-propenal Alternative name for **acrolein.**

■ **propene** $CH_3CH=CH_2$ Colourless gaseous **alkene** (olefin), used in industry for the preparation of **isopropanol, glycerol, polypropene,** etc. Alternative name: propylene.

propenoic acid Alternative name for **acrylic acid.**

propenonitrile Alternative name for **acrylonitrile.**

propionaldehyde Alternative name for **propanal**.

propionic acid Alternative name for **propanoic acid**.

protactinium Pa Radioactive element in Group IIIB of the Periodic Table (one of the **actinides**).It has several **isotopes**, with half-lives of up to 2×10^4 years. At. no. 91; r.a.m. 231 (most stable isotope).

protease **Enzyme** that breaks down **protein** into its constituent **peptides** and **amino acids** by splitting peptide linkages (*e.g.* pepsin, trypsin).

■ **protein** Member of a class of high molecular weight **polymers** composed of a variety of **amino acids** joined by **peptide** linkages. In conjugated proteins, the amino acids are joined to other groups. Proteins are extremely important in the physiological structure and functioning of all living organisms. [3/7/b]

■ **proton** Fundamental **elementary particle** with a positive charge equal in magnitude to the negative charge on an **electron**, and with a mass about 1,850 times that of an electron. Protons are constituents of the **nucleus** in every kind of **atom**. [8/8/a]

protonic acid Compound that releases solvated **hydrogen ions** in a suitable polar solvent (*e.g.* water).

proton number Alternative name for **atomic number**

prussic acid Alternative name for **hydrocyanic acid**.

pseudo- Prefix meaning false (*e.g.* pseudohalogen).

pseudohalogen Member of a group of volatile chemical compounds that chemically resemble the halogens; *e.g.* cyanogen $(CN)_2$, thiocyanogen $(SCN)_2$.

psi particle **Meson** that has no charge, but a very long lifetime. Alternative name: J-particle.

■ **PTFE** Abbreviation of **polytetrafluoroethene.**

***p*-type conductivity** Conductivity that results from the movement of positive holes (**lattice** sites of a crystalline **semiconductor** that are occupied by an acceptor impurity atom).

purine $C_5H_4N_4$ Nitrogen-containing **base** from which the bases characteristic of nucleotides and **DNA** are derived; *e.g.* adenine, guanine. Other purine derivatives include caffeine and uric acid.

putty powder Impure **tin(IV) oxide** (SnO_2), used for polishing glass.

■ **PVC** Abbreviation of **polyvinyl chloride** (polychloroethene).

pyranose Any of a group of **monosaccharide** sugars (hexoses) whose molecules have a six-membered heterocyclic ring of five carbon atoms and one oxygen atom. *See also* **furanose.**

pyrazine $C_4H_4N_2$ Heterocyclic **aromatic compound** whose ring contains four carbon atoms and two nitrogen atoms. Alternative name: 1,4-diazine.

pyrazole $C_3H_4N_2$ Heterocyclic **aromatic compound** whose ring contains three carbon atoms and two nitrogen atoms. Alternative name: 1,2-diazole.

pyrene $C_{16}H_{10}$ **Aromatic compound** consisting of four **benzene** rings fused together.

■ **Pyrex** Trade name for a heat-resistant glass, used for domestic and laboratory glassware.

pyridine C_5H_5N Heterocyclic liquid organic **base** which occurs in the light oil fraction of coal-tar and in bone oil. It forms salts with acids and is important in organic synthesis.

pyridoxine Crystalline substance from which the active **coenzyme** forms of **vitamin** B_6 are derived. It is also utilized as a potent growth factor for bacteria.

pyrimidine $C_4H_4N_2$ Heterocyclic crystalline organic **base** from which bases found in DNA are derived; its derivatives also include barbituric acid and the barbiturate drugs.

pyro- Prefix that denotes strong heat, fire.

pyrolysis Decomposition of a chemical compound by heat.

pyrometallurgy Metallurgy involved in the winning and refining of metals where heat is used, as in roasting and smelting.

pyrophoric alloy Alloy that catches fire when struck or subjected to friction, used for cigarette lighter flints.

pyrrole $(CH)_4NH$ Heterocyclic liquid organic compound, aromatic in character, whose derivatives are important biologically; *e.g.* haem, chlorophyll.

pyruvate **Ester** or **salt** of **pyruvic acid**.

pyruvic acid $CH_3COCOOH$ Simplest keto-acid, important in making energy available from ingested food. Alternative name: 2-oxopropanoic acid.

Q

■ **quadrivalent** Having a **valence** of four. Alternative name: tetravalent.

■ **qualitative** Dealing with the identity, qualities or appearance of something only.

■ **qualitative analysis** Identification of the constituents of a substance or mixture, irrespective of their amount.

quanta Plural of **quantum**.

■ **quantitative** Dealing with quantities of substances, *e.g.* mass, volume, etc., irrespective of their identity.

■ **quantitative analysis** Determination of the amounts of constituent substances present, often by weighing or manipulating volumes of solutions. *See also* **gravimetric analysis; volumetric analysis**.

quantum Unit quantity (an indivisible 'packet') of energy postulated in the **quantum theory**. The photon is the quantum of electromagnetic radiation (such as light) and in certain contexts the **meson** is the quantum of the nuclear field.

quantum mechanics Method of dealing with the behaviour of small particles such as **electrons** and **nuclei**. It uses the idea of the particle-wave duality of matter. Thus an electron has a dual nature, particle and wave, but it behaves as one or the other according to the nature of the experiment.

quantum number Integer or half-integral number that specifies possible values of a quantitized physical quantity, *e.g.* energy level, nuclear spin, angular momentum, etc.

quantum state State of an **atom, electron, particle**, etc., defined by a unique set of **quantum numbers**.

quantum theory Theory of radiation. It states that radiant energy is given out by a radiating body in separate units of energy known as **quanta**; the same applies to the absorption of radiation. The total amount of radiant energy given out or absorbed is always a whole number of quanta.

quark Subatomic particle that combines with others to form a **hadron**. Current theory predicts six types of quarks and six antiquarks, but none has yet been observed.

■ **quartz** SiO_2 Natural crystalline **silica** (silicon dioxide), one of the hardest of common minerals. Its crystals (which can generate piezoelectricity) are frequently colourless and transparent. It is used as an abrasive and in mortar and cement. [9/3/d]

quaternary ammonium compound Member of a group of white crystalline solids, soluble in water, and completely dissociated in solution. These compounds have the general formula $R_4N^+X^-$, where R is a long-chain **alkyl group**. They have **detergent** properties. Alternative name: quaternary ammonium salt.

■ **quicklime** CaO Whitish powder prepared by roasting **limestone**, used in agriculture and in cements and mortar. Alternative names: calcium oxide, lime.

quinine Colourless crystalline **alkaloid**, obtained from the bark of the cinchona shrub, once much used in the treatment and prevention of malaria.

quinoline C_9H_7N Colourless oily liquid **heterocyclic base**. It is present in coal-tar and bone oil, and was first obtained from **quinine** by alkaline decomposition. It is used in making dyes and drugs.

quinone Member of a group of cyclic **unsaturated** diketones in which the double bonds and keto groups are **conjugated**. Thus they are not **aromatic compounds**.

R

racemic acid **Racemic mixture** of **tartaric acid**.

racemic mixture Optically inactive mixture that contains equal amounts of **dextrorotatory** and **laevorotatory** forms of an **optically active** compound.

racemization Transformation of **optically active** compounds into **racemic mixtures**. It can be effected by the action of heat or light, or by the use of chemical reagents.

■ **radiation** Energy that travels in the form of electromagnetic radiation, *e.g.* radio waves, infra-red radiation, light, ultraviolet radiation, X-rays and gamma rays. The term is also applied to the rays of **alpha** and **beta particles** emitted by **radioactive** substances. Particle rays and short-wavelength electromagnetic radiation may be harmful to tissues as they are types of **ionizing radiation**.

■ **radical** Group of **atoms** within a molecule that maintains its identity through chemical changes that affect the rest of the molecule, but is usually incapable of independent existence; *e.g.* an **alkyl group** such as methyl, CH_3-. *See also* **free radical**.

■ **radioactive** Possessing or exhibiting **radioactivity**.

■ **radioactive decay** Way in which a **radioactive** element spontaneously changes into another element or **isotope** by the emission of **alpha** or **beta particles** or **gamma-rays**. The rate at which it does so is represented by its **half-life**. [6/9/d]

radioactive equilibrium Condition attained when a parent **radioactive** element produces a daughter radioactive element

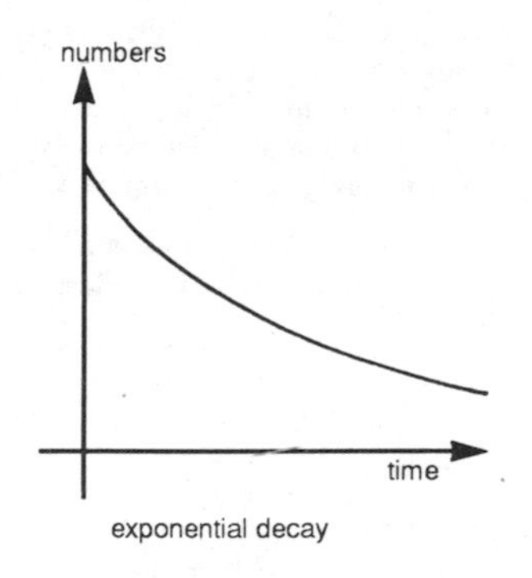

Radioactive substances decay exponentially

that decays at the same rate as it is being formed from the parent.

radioactive series One of three series that describe the **radioactive decay** of 40 or more naturally occurring radioactive **isotopes** of high **atomic number**. They are known (after the element at the beginning of each sequence) as the **thorium** series, **uranium** series and **actinium** series.

radioactive standard Radio-isotope of known rate of **radioactive decay** used for the calibration of **radiation**-measuring instruments.

■ **radioactive tracing** Use of **radio-isotopes** to study the movement and behaviour of an element through a biological or chemical system by observing the intensity of its **radioactivity**. [8/7/d]

■ **radioactive waste** Hazardous **radio-isotopes** (fission products) that accumulate as waste products in a **nuclear reactor**. They have to be periodically removed and stored safely or reprocessed. The term is also applied to the waste ('tailings') produced by the processing of uranium ores.

■ **radioactivity** Spontaneous disintegration of atomic **nuclei,** usually with the emission of **alpha particles, beta particles** or **gamma-rays**. [8/7/d]

■ **radiocarbon dating** Method of estimating the ages of carbon-containing (*e.g.* wooden) archaeological and geological specimens that are up to 50,000 years old. A **radio-isotope** of carbon, carbon-14 (C-14), is present in **carbon dioxide** and becomes assimilated into plants during photosynthesis (and into animals that eat plants). The C-14 present in 'dead' carbonaceous materials decays and is not replaced (*see* **radioactive decay**). By comparing the **radioactivities** of the 'dead' and 'live' materials, the age of the former can be estimated, because the **half-life** of C-14 is known. Alternative name: radioactive dating. [8/7/d]

■ **radiochemistry Chemistry** of **radioactive** elements and their compounds.

■ **radio-isotope Isotope** that emits **radioactivity** (ionizing radiation) during its spontaneous decay. Radio-isotopes are useful sources of radiation (*e.g.* in radiography) and are used as tracers (for **radioactive tracing**).

■ **radiology** Study of **X-rays, gamma-rays** and **radioactivity** (including **radio-isotopes**), especially as used in medicine.

radio-opaque Resistant to the penetrating effects of radiation, especially **X-rays**; often used to describe substances injected into the body before a radiography examination.

radium Ra Silver-white radioactive metallic element in Group IIA of the Periodic Table (the **alkaline earths**). It has several **isotopes**, with half-lives of up to 1,620 years. It is obtained from pitchblende (its principal ore), and used in **radiotherapy** and luminous paints. At. no. 88; r.a.m. 226 (most stable isotope).

■ **radon** Rn Radioactive gaseous element in Group 0 of the Periodic Table (the **rare gases**), a **radioactive decay** product of **radium**. It has several **isotopes**, with half-lives of up to 3.82 days. Radon coming out of the ground, particularly in hard-rock areas, is a source of background radiation that has been recognized as a health hazard. At. no. 86; r.a.m. 222 (most stable isotope). [6/9/a]

raffinate Liquid that remains after a substance has been obtained by **solvent extraction**.

raffinose $C_{18}H_{32}O_{16}$ Colourless crystalline **trisaccharide carbohydrate** that occurs in sugar beet, which hydrolyses to the **sugars** galactose, glucose and fructose.

■ **r.a.m.** Abbreviation of **relative atomic mass** (formerly called atomic weight).

Raney nickel Spongy form of **nickel**, made by treating an aluminium-nickel alloy with sodium hydroxide solution, which is used as a **catalyst**, particularly in the **hydrogenation** of **fats and oils**. It was named after the American chemist M. Raney.

Raoult's law The relative lowering of the **vapour pressure** of a **solution** is proportional to the **mole** fraction of the **solute** in the solution at a particular temperature. It was named after the French scientist François Raoult (1830–1901).

rare-earth element Alternative name for a member of the series of elements, in Group IIIB of the **Periodic Table**, known as the **lanthanides**.

■ **rare gas** One of the uncommom, unreactive and highly stable gases in Group 0 of the Periodic Table. They are **helium, neon, argon, krypton, xenon** and **radon**. Alternative names: inert gas, noble gas. [6/9/a]

rate constant Constant of proportionality for the speed of a **chemical reaction** at a particular temperature. It can only be obtained experimentally. Alternative names: velocity constant, specific rate constant.

rate-determining step Slowest step of a **chemical reaction** which determines the overall rate, provided the other steps are relatively rapid. Thus the kinetics and **order of reaction** are basically those of the rate-determining step. Alternative name: limiting step.

rate equation Alternative name for **rate law**.

rate law Equation that relates the rate of a **chemical reaction** to the **concentration** of the individual reactants. It has the form rate $= k[X]^n$, where k is the **rate constant**, X is the **reactant** and n is the **order of reaction**. It can only be obtained experimentally. Alternative name: rate equation.

■ **rate of reaction** Speed of a **chemical reaction**, usually expressed as the change in **concentration** of a reactant or product per unit time. It can be affected by temperature, pressure and the presence of a **catalyst**. [7/7/a]

■ **raw material** Substance from which others are made. It may be simple or complex. *E.g.* nitrogen (from air) and hydrogen are raw materials for the synthesis of ammonia; coal and petroleum are raw materials from which a wide range of complex chemicals are made.

■ **rayon** Man-made fibre manufactured from **cellulose** (obtained from wood pulp), used for making textiles. For acetate rayon, a solution of cellulose acetate in a volatile solvent is extruded through spinnerets into air. For viscose rayon, wood pulp is dissolved in sodium hydroxide and carbon disulphide to produce cellulose xanthate, which is extruded through spinnerets into a bath of sulphuric acid to re-form cellulose fibres. Industrial uses of rayon include conveyor belts and hoses. Former name: artificial silk.

■ **reactant** Substance that reacts with another in a **chemical reaction** to form new substance(s).

■ **reaction** Alternative name for **chemical reaction**.

reactive dye Dye that forms a **covalent bond** with the fibre molecule of the textile being dyed. This provides excellent fastness. Such dyes are used to dye cellulose fibres (*e.g.* rayon).

reagent *1*. Substance that takes part in a **chemical reaction**; a reactant. *2*. Common laboratory chemical used in chemical analysis and for experiments.

realgar As_2S_2 Natural red arsenic disulphide, used as a pigment and in pyrotechnics.

real gas Gas that never fully achieves 'ideal' behaviour. *See also* **ideal gas**.

reciprocal proportions, law of Alternative name for the law of **equivalent proportions**.

■ **recrystallization** *1*. Change from one crystal structure to another; it occurs on heating or cooling through a critical temperature. *2*. Purification of a substance by repeated **crystallization** from solution.

rectification Purification of a liquid using **distillation**.

■ **rectified spirit** Solution of **ethanol** (ethyl alcohol) that contains about 5–7% water. It is a constant-boiling mixture and the water cannot be removed by **distillation**.

■ **recycling** Method of conserving resources that involves sorting waste materials into their chief components (metals, plastic, paper and glass) and using them to make more of the component materials. *E.g.* up to half the aluminium produced comes from recycled soft drinks cans. [2/6/b]

red clay Fine-grained sediment, covering about 50% of the mid-ocean floors, formed from material that originated on land and was carried far out to sea by rivers and the wind.

■ **red lead** Pb_3O_4 Bright red powdery oxide of **lead**. An **oxidizing agent**, it is used in anti-rust and priming paints. Alternative names: minium, dilead(II) lead(IV) oxide, lead tetraoxide, triplumbic tetroxide.

■ **redox reaction Chemical reaction** in which **oxidation** is necessarily accompanied by **reduction**, and vice versa; an oxidation-reduction reaction. [7/7/b]

reduced equation of state Law which states that if any two or more substances have the same reduced pressure π, *i.e.* their pressures are the same fraction or multiple π of their respective critical pressures, and are at equal reduced temperatures θ, then their reduced volumes ϕ should be equal.

reduced pressure distillation Alternative name for **vacuum distillation**.

reduced temperature, pressure and volume Quantities θ, π and ϕ which are the ratios of the temperature, pressure and volume to the **critical temperature, critical pressure** and **critical volume** respectively in the **reduced equation of state**.

■ **reducing agent** Substance that causes chemical **reduction**, often by adding hydrogen or removing oxygen; *e.g.* carbon, carbon monoxide, hydrogen. Alternative name: electron donor. [7/7/b]

reducing sugar Any **sugar** that can act as a **reducing agent**. *See also* **Benedict's test**; **Fehling's test**.

reductase Enzyme that causes the **reduction** of an organic compound.

■ **reduction** Chemical reaction that involves the gain of **electrons** by a substance; the addition of hydrogen or removal of **oxygen** from a substance. [7/7/b]

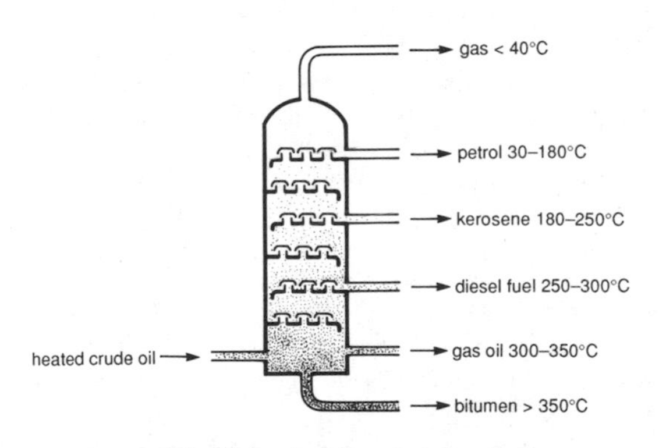

Oil refining involves fractionation

■ **refining** *1.* Purification; the removal of impurities from a substance, particularly a crude metal after it has been extracted from its ore. *2.* Splitting of **petroleum** into its component hydrocarbons, usually by **fractional distillation**. [7/9/a]

■ **reflux** Boiling of a liquid for long periods of time. Loss by evaporation is prevented by using a **reflux condenser**.

■ **reflux condenser** Vertical condenser used in the process of refluxing. It is attached to a vessel that contains the liquid to be refluxed and condenses the vapour produced on boiling, which then runs back into the vessel.

reforming Production of branched-chain **alkanes** from straight-chain ones or the production of **aromatic compounds** (*e.g.* benzene) from **alkenes**, using **cracking** or a **catalyst**.

■ **refrigerant** Substance used as the working fluid in a refrigerator (*e.g.* ammonia, fluon, CFCs).

■ **relative atomic mass** (r.a.m.) Mass of an atom relative to the mass of the isotope carbon-12 (which is taken to be exactly 12). Former name: atomic weight.

■ **relative molecular mass** Alternative name for molecular weight.

■ **renewable resources** Natural material that can be renewed (*e.g.* timber can eventually be replaced by planting more trees). **Fossil fuels** are not renewable because they cannot be replaced once they have been used up. [2/6/b]

renin Enzyme produced by the kidney that constricts arteries and thus raises blood pressure.

rennin Enzyme found in gastric juice that curdles milk. It is the active ingredient of rennet.

replicase Enzyme that promotes the synthesis of **DNA** and **RNA** within living cells.

resin Organic compound that is generally a viscous liquid or semi-liquid which gradually hardens when exposed to air, becoming an amorphous, brittle solid. Natural resins, found in plants, are yellowish in colour and insoluble in water, but are quite soluble in organic solvents. Synthetic resins (types of plastics) also possess many of these properties. *See also* **rosin**.

resonance Movement of **electrons** from one **atom** of a **molecule** or **ion** to another atom of that molecule or ion.

resorcinol $C_6H_4(OH)_2$ Crystalline dihydric **phenol**, used in the synthesis of drugs and dyes. Alternative names: *m*-dihydroxybenzene, 1,3 – benzenediol.

■ **respiration** Release of energy by living organisms from the breakdown of organic compounds. In **aerobic** respiration, which occurs in most cells, **oxygen** is required and **carbon dioxide** and water are produced. Energy production is coupled to a series of **oxidation-reduction reactions**, catalysed by **enzymes**. In **anaerobic** respiration (*e.g.* **fermentation**), food substances are only partly broken down, and thus less energy is released and oxygen is not required. [3/6/a]

■ **respiratory pigment** Substance that can take up and carry **oxygen** in areas of high oxygen concentration, releasing it in parts of the organism with low oxygen concentrations where it is consumed, *i.e.* by **respiration** in cells. In vertebrates the respiratory pigment is haemoglobin; in some invertebrates it is haemocyanin.

■ **retort** Heated vessel used for the **distillation** of substances, as in the separation of some metals.

reverberatory furnace Furnace for **smelting** metal **ores** in which the flame and hot gases are reflected downwards by a curved roof to the material to be heated. The fuel is in one part of the furnace and the ore in another.

■ **reversible process** Process that can theoretically be reversed by an appropriate small change in any of the thermodynamic variables (*e.g.* pressure, temperature). Real natural processes are irreversible.

■ **reversible reaction Chemical reaction** that can go either forwards or backwards depending on the conditions.

rhenium Re Rare metallic element of Group VIIB of the Periodic Table (a **transition element**), used in making thermocouples. At no 75; r.a.m. 186.20.

rhodium Rh Silver-white metallic element in Group VIII of the Periodic Table (a **transition element**), used as a **catalyst** and in making thermocouples. At. no. 45; r.a.m. 102.905.

■ **riboflavin** Orange water-soluble crystalline solid, member of the vitamin B complex. It plays an important role in growth. Alternative names: riboflavine, lactoflavin, vitamin B_2. [3/9/b]

ribonucleic acid *See* **RNA**.

ribose $C_5H_{10}O_5$ Optically active **monosaccharide** sugar, a component of the nucleotides of **RNA**.

RNA Abbreviation of ribonucleic acid, one of the nucleic acids present in cells, the other being **DNA**. It is composed of nucleotides that contain **ribose** as the sugar. Messenger RNA takes part in transcription or copying of the genetic code from a DNA template. Transfer RNA and ribosomal RNA take part in translation or protein synthesis.

■ **roasting** Heating a compound (*e.g.* a metal ore) in air to convert it to an **oxide**.

■ **rock** Mixture of minerals; any distinguishable portion of the Earth's crust, soft or hard, loose or consolidated, *e.g.* granite, limestone, clay, coal and sand. [7/10/c]

■ **rock salt** Naturally occurring crystalline **sodium chloride**, an important raw material in the chemical industry.

rosin Amber brittle translucent **resin** obtained as a residue after distilling **turpentine**. It is used in making paints and varnishes, as a flux for soldering, and (as a powder) to give grip to boxers' shoes and violin bows.

■ **rubber** Elastic substance obtained from plant **latex**. It is a **polymer**, containing long chains of the monomer **isoprene**. Alternative name: natural rubber.

■ **rubber, synthetic** Synthetic compound with a structure resembling that of natural **rubber** in consisting of long-chain **polymer** molecules; these are built of **monomers** such as **acetylene** (ethyne), **isoprene, styrene** and **vinyl** compounds.

rubidium Rb Reactive silver-white metal in Group IA of the Periodic Table (the **alkali metals**), with a naturally occurring radioactive **isotope** (Rb-87). At. no. 37; r.a.m. 85.47.

■ **ruby** Natural red impure form of **corundum** (aluminium oxide, Al_2O_3). Imperfect rubies are used as bearings in watches etc. Synthetic rubies are used as gems or in industry.

■ **rust** Red-brown corrosion product consisting of hydrated iron oxides that forms on the surface of **iron** and **steel** exposed to moisture and air. [7/7/d]

ruthenium Ru Silver-white metallic element in Group VIII of the Periodic Table (a **transition element**), used to add hardness to platinum alloys. At. no. 44; r.a.m. 101.07.

rutherfordium Rf Element no. 104 (a **post-actinide**). It is a radioactive metal with at least three very short-lived isotopes (half-lives up to 70 seconds), made by bombarding an **actinide** with carbon, oxygen or neon atoms. Alternative name: kurchatovium.

rutile Natural form of **titanium(IV) oxide** (TiO_4) and a minor **ore** of **titanium**.

Rydberg constant (R) Constant relating to the wave number of atomic spectrum lines. Its value for **hydrogen** is 1.09677×10^7 m^{-1}. It was named after the Swedish physicist Johannes Rydberg (1854–1919).

Rydberg formula Formula for expressing the wave number of a spectral line. The general Rydberg formula is $1/\lambda = R(1/n^2 - 1/m^2)$, where n and m are positive integers, and R is the **Rydberg constant**.

S

■ **saccharide** Simplest type of **carbohydrate,** with the general formula $(C_6H_{12}O_6)_n$, common to many **sugars**. Alternative name: saccharose. *See also* **disaccharide; monosaccharide; polysaccharide.**

saccharimetry Measurement of the concentration of **sugar** in a solution from its **optical activity**, by using a **polarimeter**.

saccharin $C_6H_4SO_2CONH$ White crystalline organic compound that is about 550 times sweeter than **sugar**; an artificial sweetener. It is almost insoluble in water and hence it is used in the form of its soluble sodium salt. Alternative names: 2-sulphobenzimide, saccharine.

saccharose Alternative name for **saccharide**.

■ **sacrificial anode** Method of protecting a steel structure (*e.g.* a bridge, underground or underwater pipeline) by wiring to it plates of an electropositive metal such as magnesium. In a damp or wet environment the magnesium acts as an anode and corrodes in preference to the steel (the cathode); the anodes are periodically replaced with new ones. The technique is called sacrificial protection. [11/7/b]

■ **safety lamp** Oil lamp designed to be taken into coal mines. It has the flame surrounded by fine wire gauze, so that it can be alight in an atmosphere containing **firedamp** (methane) without causing an explosion. Alternative name: Davy lamp.

sal ammoniac Old name for **ammonium chloride**.

salicylate Ester or **salt** of **salicylic acid**.

salicylic acid $C_6H_4(OH)COOH$ White crystalline organic compound, a **carboxylic acid**. It is used as an antiseptic, in medicine, and in the preparation of **azo dyes**. Its acetyl **ester** is aspirin. Alternative name: 2-hydroxybenzoic acid.

saline Salty; describing a solution of **sodium chloride** (common salt).

■ **salt** *1.* Product obtained when a **hydrogen** atom in an **acid** is replaced by a **metal** or its equivalent (*e.g.* the ammonium ion NH_4^+). This occurs when a metal dissolves in an acid; *e.g.* zinc (Zn) dissolves in sulphuric acid (H_2SO_4) to form zinc sulphate ($ZnSO_4$) and water (H_2O). A salt is also formed when a **base** or **alkali** neutralizes an acid (*see* **neutralization**), or in a **double decomposition** reaction. *See also* **acid salt**; **basic salt**. *2.* Alternative name for common salt, **sodium chloride** (NaCl). [7/6/a]

salt bridge Tube that contains a **saturated solution** of **potassium chloride**, or an **agar** gel made with concentrated potassium chloride solution. It is employed to connect two **half-cells**.

saltcake Alternative name for crude **sodium sulphate**.

salting out **Precipitation** of a **colloid** (*e.g.* gelatine) by the addition of large amounts of a salt.

■ **saltpetre** Alternative name for **potassium nitrate**.

sal volatile Old name for **ammonium carbonate**.

samarium Sm Metallic element in Group IIIB of the Periodic Table (one of the **lanthanides**). It is slightly radioactive and arises from fission fragments in a **nuclear reactor**, where it acts as a 'poison'. At. no. 62; r.a.m. 150.35.

■ **sand** Separate grains of **quartz**; the indestructible residue from the erosion of rocks. It is used in mortar and concrete and in making glass.

■ **sandstone** Consolidated **sand**, cemented together with clay, **calcium carbonate** or iron oxide. It is used as a building stone. [9/2/d] [9/3/d]

sandwich compound Organometallic compound whose molecules consist of two parallel planar rings with a **metal** atom centred between them (*e.g.* **ferrocene**).

■ **saponification Hydrolysis** of an **ester**, using an **alkali**, to produce a free **alcohol** and a **salt** of the organic **acid**. It is the process by which **soap** is made.

saponification value Number of milligrams of **potassium hydroxide** required for the complete **saponification** of 1 g of the substance being tested. Alternative name: saponification number.

■ **sapphire** Blue gem variety of **corundum** (aluminium oxide, Al_2O_3).

saprolite Naturally occurring deposit of rock that has broken down chemically (*e.g.* by the action of acids in rainwater).

■ **saturated compound** Organic compound that contains only **single bonds**; all the atoms in the compound exert their maximum combining power (valence) with other atoms, so that a chemical change can be effected only by substitution and not by addition. *See* **addition reaction; substitution reaction.**

■ **saturated solution Solution** that cannot take up any more **solute** at a given temperature. *See also* **supersaturated solution.**

saturated vapour Vapour that can exist in equilibrium with its parent **solid** or **liquid** at a given temperature.

saturated vapour pressure Pressure exerted by a **saturated vapour**. It is temperature dependent.

■ **saturation** Point at which no more of a material can be dissolved, absorbed or retained by another.

s-**block elements** Metallic elements that form Groups IA and IIA of the Periodic Table, which include the **alkali metals, alkaline earths** and the **lanthanides** and **actinides** (together also known as the **rare earths**); hydrogen is usually included as well. They are so called because their 1 or 2 outer electrons occupy *s*-**orbitals**.

scandium Sc Silvery-white metallic element in Group IIIB of the Periodic Table (a **rare earth** element); its oxide, Sc_2O_3, is used as a catalyst and to make ceramics. At. no. 21; r.a.m. 44.956.

Schiff's base Organic compound formed when an **aldehyde** or **ketone** condenses with a primary aromatic **amine** with the elimination of water. Alternative names: aldimine, azomethine. It was named after the German chemist Hugo Schiff (1834–1915).

Schiff's reagent Reagent for testing for the presence of aliphatic **aldehydes**, which quickly restore its magenta colour. It is prepared by dissolving rosaniline hydrochloride in water and passing sulphur dioxide through it until the magenta colour is discharged.

Schrödinger equation Equation that treats the behaviour of an **electron** in an **atom** as a three-dimensional stationary wave. Its solution is related to the probability that the electron is located in a particular place. It was named after the Austrian physicist Erwin Schrödinger (1887–1961). Alternative name: Schrödinger wave equation.

■ **secondary cell Electrolytic cell** that must be supplied with electric charge before use by passing a direct current through it, but it can be recharged over and over again. Alternative names: accumulator, storage cell. *See also* **primary cell**.

sedimentation Removal of solid particles from a **suspension** by gravitational force or in a **centrifuge**.

selectively permeable membrane Alternative name for **semipermeable membrane**.

selenium Se Nonmetallic element in Group VIA of the Periodic Table, obtained from flue dust in refineries that use **sulphide** ores. One of its **allotropes** conducts electricity in the presence of light, and is used in photocells and rectifiers. At. no. 34; r.a.m. 78.96.

semicarbazide $NH_2NHCONH_2$ Organic **base** which forms salts (*e.g.* it forms a hydrochloride with hydrochloric acid) and characteristic condensation products (**semicarbazones**) with **aldehydes** and **ketones**.

semicarbazone Crystalline organic compound containing the group $=C=NNHCONH_2$, formed when an **aldehyde** or a **ketone** reacts with **semicarbazide** ($NH_2NHCONH_2$) with the elimination of water. Semicarbazones are used to identify the original aldehyde or ketone.

semiconductor Substance that is an **insulator** at very low temperatures but which becomes a **conductor** if the temperature is raised or when it is slightly impure. Semiconductors are used in rectifiers and photoelectric devices, and for making diodes and transistors. *See also* **donor**; **doping**; ***n*-type conductivity**; ***p*-type conductivity**.

semipermeable membrane Porous **membrane** that permits the passage of some substances but not others; *e.g.* a cell's plasma membrane, which permits entry of small molecules such as water but not large molecules, allowing **osmosis** to occur. Such membranes are extremely important in biological systems and are used in **dialysis**. Alternative name: selectively permeable membrane.

septivalent Having a valency of seven. Alternative name: heptavalent.

■ **series** Systematically arranged succession of chemical compounds (*e.g.* homologous series) or of numbers or algebraic terms (*e.g.* arithmetic series, exponential series, geometric series).

serine $CH_2OHCHNH_2COOH$ White crystalline **amino acid**, present in many **proteins**. Alternative name: 2-amino-3-hydroxypropanoic acid.

■ **shell** Group of **electrons** that share the same principal **quantum number** in an atom. [8/8/c]

side chain **Aliphatic** radical or group that is attached to one of the atoms in the ring of a **cyclic** compound or to one of the atoms of the longer straight chain of atoms in an **acyclic** compound.

side reaction **Chemical reaction** that takes place at the same time as the main reaction.

■ **Siemens-Martin process** Alternative name for **open-hearth process**.

silane **Hydride** of **silicon** with the general formula Si_nH_{2n+2}. The maximum number of bonds possible between silicon atoms is about eight because of the weakness of such bonds. Alternative names: silicon hydride, silicane.

■ **silica** SiO_2 Hard white solid with a high melting point (1,710°C). It is one of the most abundant natural compounds, occurring as *e.g.* crystobalite, **flint**, **quartz**, **sand**, and combined as **silicates** in rocks. Alternative name: silicon(IV) oxide, silicon dioxide. *See also* **silica gel**. [9/2/d]

■ **silica gel** Porous **amorphous** variety of **silica** (SiO_2) which is capable of absorbing large quantities of water and other **solvents**. It is used as a **desiccant** and **adsorbent**. [9/3/d]

silicate Derivative of **silica** or a salt of a **silicic acid**. Natural silicates, found in many rocks and minerals, contain varying proportions of silica and of a wide range of metal oxides. *See* **glass; sodium silicate**.

silicic acid Hydrated form of **silica** (SiO_2) made by reacting a soluble **silicate** with an **acid**. It forms a colloidal or gel-like mass.

silicon Si Nonmetallic element in Group IVA of the Periodic Table, which exists as amorphous and crystalline **allotropes**. It is the second most abundant element, occurring as **silicates** in clays and rocks; sand and quartz consists of **silica** (silicon dioxide, SiO_2). It is used in making refractory materials and temperature-resistant glass. At. no. 14; r.a.m. 28.086.

silicon carbide SiC Very hard black material, used as an abrasive, and in cutting, grinding and polishing instruments. Alternative name: carborundum.

silicon(IV) chloride Alternative name for **silicon tetrachloride**.

silicon dioxide Alternative name for **silica**.

silicone Member of a group of synthetic **polymers** made from **siloxanes**, used in textile finishing, polishes and lubricants.

silicon(IV) oxide Alternative name for **silica**.

silicon tetrachloride $SiCl_4$ Colourless fuming liquid, a source of pure **silica** for use in the production of silica **glass**. Alternative name: silicon(IV) chloride.

siloxane Member of a group of compounds that contain the linkage Si-O-Si with organic groups bound to the **silicon** atoms. They are used for making **silicones**.

■ **silver** Ag Silver-white metallic element in Group IB of the Periodic Table (a **transition element**). It occurs as the free element (native) and in various **sulphide** ores. It is used in jewellery, electrical contacts, batteries and mirrors. Silver **halides** are used in photographic emulsions. At. no. 47; r.a.m. 107.868.

silver bromide AgBr Pale yellow insoluble crystalline salt, used for making light-sensitive photographic emulsions.

silver chloride AgCl White insoluble crystalline salt, used in the manufacture of pure **silver** and in photographic emulsions.

silver iodide AgI Pale yellow insoluble crystalline salt, used in photographic emulsions.

■ **silver nitrate** $AgNO_3$ Colourless crystalline salt, used in **volumetric analysis** and **silver plating**, and as a caustic in medicine (*e.g.* for removing warts).

silver oxide Ag_2O Brown amorphous solid, only slightly soluble in water but soluble in ammonia solution. Alternative name: silver(I) oxide.

silver plating Electrolytic deposition of a layer of **silver** on the surface of another metal, usually from a hot alkaline solution of complex silver(I) cyanides. Alternative name: silvering.

■ **single bond Covalent bond** formed by the sharing of one pair of **electrons** between two **atoms**. [8/8/c]

sintering Method of compacting a powdered solid into a rigid shape by compressing it at a temperature below its melting point.

■ **SI units** Abbreviation for Système International d'Unités, an international system of scientific units. It has seven basic units: metre (m), kilogram (kg), second (s), kelvin (K), ampere (A), mole (mol) and candela (cd), and two supplementary units radian (rad) and steradian (sr).

■ **slag** Impurities, mainly silicates, that rise to the surface during the **smelting** of a metal ore. [7/7/c]

■ **slaked lime** Alternative name for **calcium hydroxide**.

slow neutron Alternative name for **thermal neutron**.

■ **smelt** Material obtained from the **smelting** of a metal ore.

■ **smelting** Thermal decomposition of concentrated metal ore to cause the release of the metal. [7/7/c]

■ **smoke** Particles of a **solid** dispersed in a **gas**.

■ **soap** Sodium or potassium salt of a high molecular weight **fatty acid** (*e.g.* palmitic acid, stearic acid). Soaps are made by the **hydrolysis** or **saponification** of **fats** with hot **sodium hydroxide** or **potassium hydroxide**, giving **glycerol** as a by-product. They emulsify grease and act as wetting agents. *See also* **detergent**.

soapstone Compact form of **chalk**. Alternative name: steatite.

■ **soda** Imprecise term for a compound of **sodium**, usually referring to **sodium carbonate** (washing soda) or **sodium hydrogencarbonate** (baking soda). *See also* **caustic soda**; **soda ash**; **soda lime**.

■ **soda ash** Common name for anhydrous **sodium carbonate**.

■ **soda lime** Solid mixture of **sodium hydroxide** and **calcium oxide**.

■ **soda water** **Carbon dioxide** dissolved in **water** under pressure, used as a fizzy drink.

■ **sodium** Na Soft silvery-white metallic element in Group IA of the Periodic Table (the **alkali metals**). It occurs widely, principally as its **chloride** (common salt, NaCl) in seawater and as underground deposits, from which it is extracted by **electrolysis**. The metal is used as a coolant in some **nuclear reactors**; its many compounds are important in the chemical industry, particularly – in addition to the chloride – **sodium hydroxide** (caustic soda, NaOH) and **sodium carbonate** (soda, Na_2CO_3). At. no. 11; r.a.m. 22.9898.

sodium acetate CH_3COONa White crystalline solid, used in photography and in the manufacture of **ethyl ethanoate** (acetate) and various **pigments**. Alternative name: sodium ethanoate.

sodium aluminate $NaAlO_2$ White solid, used as a coagulant in water purification.

sodium azide NaN_3 White poisonous crystalline solid, used in the manufacture of detonators.

■ **sodium bicarbonate** Alternative name for **sodium hydrogencarbonate**.

sodium bisulphate Alternative name for **sodium hydrogensulphate**.

sodium bisulphite Alternative name for **sodium hydrogensulphite**.

sodium borate Alternative name for **borax**.

sodium bromide NaBr White crystalline solid, used in medicine.

■ **sodium carbonate** $Na_2CO_3.10H_2O$ White crystalline solid which exhibits **efflorescence** and forms an alkaline solution in water. It is used in **glass** making, as a water softener, and for the preparation of **sodium** chemicals. Alternative names: washing soda, soda, soda ash.

sodium chlorate $NaClO_3$ White soluble crystalline solid. It is a powerful **oxidizing agent**, used as a weed-killer and in the textile industry. Alternative name: sodium chlorate(V).

■ **sodium chloride** NaCl White soluble crystalline salt, extracted from seawater or underground deposits. It is used for seasoning and preserving food. Industrially, it is used in the manufacture of a wide variety of chemicals, including chlorine, sodium carbonate, sodium hydroxide and hydrochloric acid. Alternative names: common salt, rock salt, salt, sea salt, table salt.

sodium dihydrogenphosphate(V) NaH_2PO_4 White solid, used in detergents and certain baking powders. Alternative name: sodium dihydrogen orthophosphate.

sodium ethanoate Alternative name for **sodium acetate**.

■ **sodium hydrogencarbonate** $NaHCO_3$ White soluble powder, used in making baking powder, powder-based fire extinguishers and antacids. Alternative names: sodium bicarbonate, baking soda.

sodium hydrogensulphate $NaHSO_4.H_2O$ White solid, used in the dyeing industry and in the manufacture of **sulphuric acid**. Alternative name: sodium bisulphate.

sodium hydrogensulphite $NaHSO_3$ White powder, used in medicine as an antiseptic, and as a preservative. Alternative name: sodium bisulphite.

■ **sodium hydroxide** NaOH White deliquescent solid; a strong **base**. It is made by the **electrolysis** of brine (**sodium chloride** solution). It is used in the manufacture of **soaps**, **rayon** and paper and many other **sodium** compounds. Alternative names: caustic soda, soda.

sodium metasilicate Alternative name for **sodium silicate**.

sodium monoxide Na_2O White solid that reacts violently with water to give **sodium hydroxide**; a **strong base**. Alternative name: disodium oxide.

■ **sodium nitrate** $NaNO_3$ White crystalline salt, used as a food preservative and in the manufacture of explosives and fireworks. Alternative names: Chilean saltpetre, soda nitre.

sodium peroxide Na_2O_2 Pale yellow powdery solid that reacts readily with water to give **sodium hydroxide** and **oxygen**. It is an **oxidizing agent**, used as a bleach (for cloth and wood pulp).

sodium silicate $Na_2SiO_3.5H_2O$ Colourless crystalline solid, used in various types of **detergents** and cleaning compounds, and as a bonding agent in many ceramic cements and in various refractory applications. Alternative name: sodium metasilicate.

sodium silicate solution Concentrated **solution** of **sodium silicate** in water, used to prepare **silica gel** and precipitated silica. Alternative name: water glass.

sodium sulphate $NaSO_4.10H_2O$ White crystalline salt, used in the manufacture of paper, glass, dyes and **detergents**. Alternative name: Glauber's salt, saltcake.

sodium sulphide NaS_2 Reddish-yellow deliquescent amorphous solid, used in the manufacture of dyes.

sodium sulphite Na_2SO_3 White soluble crystalline solid which reacts with an acid to produce **sulphur dioxide** (SO_2), used in photography and as a bleach and food preservative.

■ **sodium thiosulphate** $Na_2S_2O_3$ White soluble crystalline solid. It is a strong **reducing agent**, used as photographic fixing agent (when it reacts with unexposed silver halides) and in dyeing and **volumetric analysis**. Alternative name: hypo.

■ **softening of water** *See* **hardness of water; ion exchange**. [5/6/a]

■ **soft iron Iron** that has a low content of **carbon**, unlike **steel**. It is unable to retain magnetism.

■ **soft water Water** that lathers immediately with **soap**; water from which most of the **calcium** and **magnesium** compounds have been removed. *See* **hardness of water**. [5/6/a]

■ **solar energy** Energy derived from the Sun. Sunlight may be used with a photocell to generate electricity, or the Sun's rays may heat water in a radiator or, focused by mirrors, work a solar furnace. [13/6/d]

■ **solder** Low-melting-point **alloy**, often containing tin and lead, used for joining various metals (with the aid of a **flux**). It melts readily and remains molten long enough for the joint to be shaped. Alternative name: braze.

■ **solid** Substance in the **solid state**.

solid solution Single solid **homogeneous** crystalline **phase** of two or more substances. Many **alloys** are solid solutions.

■ **solid state** Physical state of matter that has a definite shape and resists having it changed; the volume of a solid changes only slightly with temperature and pressure. A true solid state is associated with a definite **crystalline** form, although **amorphous** solids also exist (*e.g.* glass). Alternative name: solid. [8/6/a]

■ **solubility** Amount of a substance (**solute**) that will dissolve in a liquid (**solvent**) at a given temperature, usually expressed as a weight per unit volume (*e.g.* gm per litre) or as a percentage. *See also* **concentration**. [6/4/a]

solubility product When a **solution** is saturated with an **electrolyte**, the solubility product is the product of the **concentrations** of its constituent **ions**.

■ **soluble** Describing a substance (**solute**) that will dissolve in a liquid (**solvent**).

■ **solute** Substance that dissolves in a **solvent** to form a **solution**.

■ **solution** Homogeneous mixture of a **solute** and **solvent**.

solvation Attachment between **solvent** and **solute** molecules. The greater the polarity of the solvent, the greater is the attraction between solute and solvent molecules. It is what causes ionic compounds to dissolve, and water is a good polar solvent. Solvation by water is called **hydration**.

■ **Solvay process** Process for the commercial preparation of **sodium carbonate** (by reacting sodium chloride solution containing ammonia with carbon dioxide made by heating calcium carbonate). It was named after the Belgian chemist Ernest Solvay (1838–1922). Alternative name: ammonia-soda process.

■ **solvent** Substance (usually a liquid) in which a **solute** dissolves; the component of a **solution** which is in excess.

solvent extraction Removal of a substance from a (usually aqueous) solution by dissolving it in a (usually organic) **solvent**. Alternative name: liquid-liquid extraction.

solvent naphtha Alternative name for **naphtha**.

solvolysis Chemical reaction between **solvent** and **solute** molecules. *See also* **hydrolysis**.

sorbitol Hexahydric **alcohol**, an **isomer** of **mannitol**, formed by the **reduction** of **glucose**, used as a sweetening agent.

spartalite Alternative name for **zincite**.

specific activity Number of disintegrations of a **radio-isotope** per unit time per unit mass.

specific charge Ratio of electric charge to unit mass of an **elementary particle**.

spectrochemistry Branch of **chemistry** concerned with the study of spectra of substances.

spectrometer Spectroscope that has some form of photographic or electrical detection device.

spectrometry Measurement of the intensity of spectral lines or spectral series as a function of wavelength, using a **spectrometer** or **spectroscope**. It is an important method of chemical analysis.

spectroscope Instrument for splitting various wavelengths of electromagnetic radiation into a spectrum, using a prism or diffraction grating.

spectroscopy Study of the properties of light, using a **spectroscope**; the production and analysis of spectra.

spin, electron Spinning motion of an **electron** in an **atom**.

■ **spirit** *1.* Volatile liquid obtained by **distillation**; a volatile distillate (*e.g.* aviation spirit). *2.* Solution that consists of a volatile substance dissolved in **ethanol** (ethyl alcohol). *See also* **methylated spirits**.

■ **spontaneous combustion** Self-ignition of a substance of low **ignition temperature** (*e.g.* damp straw or oily rags), which results from the accumulation of heat within the substance because of slow **oxidation**.

■ **stable** Relatively inert; hard to decompose.

■ **stainless steel** Member of a group of **chromium**-containing **steels** characterized by a high degree of resistance to corrosion. They are used for cutlery, chemical plant equipment, ball-bearings, etc.

■ **stalactite** Long icicle-like formation of **calcium carbonate** (from dripping water containing dissolved calcium carbonate) which grows very slowly downwards from the roof of a limestone cave. *See also* **stalagmite**.

■ **stalagmite** Formation similar to a **stalactite**, but which grows upwards from the floor of a cave.

standard cell **Electrolytic cell** characterized by having a known constant electromotive force (e.m.f.).

standard electrode **Electrode** (usually a **hydrogen electrode**) that is used as a standard for measuring **electrode potentials**. It is by convention assigned a potential of zero.

standard electrode potential **Electrode potential** specified by comparison with a **standard electrode**.

standard solution **Solution** of definite **concentration**, *i.e.* having a known weight of **solute** in a definite volume of solution, as used in **volumetric analysis**. The concentration is commonly given in terms of normality (*see* **normal**).

standard state Element in its most stable physical form at a specified temperature and standard pressure (101,325 pascals or 760 mm Hg).

■ **standard temperature and pressure** (STP or s.t.p.) Set of standard conditions of temperature and pressure. By convention, the standard temperature is 273.16 K (0°C) and the standard pressure is 101,325 pascals (760 mm Hg).

stannane Alternative name for **tin(IV) hydride**.

stannic Tetravalent tin. Alternative names: tin(IV), tin(4^+).

stannic acid H_2SnO_3 White amorphous solid, used for polishing glass and metal. Alternative name: tin hydroxide oxide.

stannic chloride Alternative name for **tin(IV) chloride**.

stannous Bivalent tin. Alternative names: tin(II), tin (2^+).

stannous chloride Alternative name for **tin(II) chloride**.

■ **starch** $(C_6H_{10}O_5)_n$ Complex **polysaccharide carbohydrate**, a **polymer** of **glucose**, that occurs in all green plants, where it serves as a reserve energy material (*e.g.* in roots, tubers and cereal seeds). It forms glucose on complete **hydrolysis**. Alternative name: amylum. [3/6/b]

statistical mechanics Branch of statistics concerned with the study of the properties and behaviour of the component microscopic particles of a macroscopic system.

■ **steam Water** (H_2O) in the vapour or gaseous state; water above its **boiling point** (100°C). The white clouds near boiling water often called steam are in fact minute water droplets that have condensed out as steam cools.

■ **steam distillation Distillation** of a substance by bubbling **steam** through the heated **liquid**. It is used to purify water-insoluble volatile substances.

■ **steam point** Normal **boiling point** of water; it is taken to be a temperature of 100°C (at normal pressure).

stearate Ester or **salt** of **stearic acid**. *See also* **soap**.

stearic acid $CH_3(CH_2)_{16}COOH$ Long-chain **fatty acid** that occurs in most **fats and oils**. Its sodium and potassium salts are constituents of **soaps**. Alternative name: octadecanoic acid.

steatite Alternative name for **soapstone**.

■ **steel Iron** containing small but controlled quantities of **carbon** (0.1–1.5%) and free from silicon, sulphur and phosphorus. Its properties depend on the percentage of carbon; the more carbon, the harder the steel. It is used to make ships, bridges, frameworks of buildings, tools and reinforced concrete, etc. *See also* **stainless steel**. [7/7/c]

stereochemistry Branch of **chemistry** concerned with the study of the spatial arrangement of the **atoms** within a molecule and the way that these affect the properties of that molecule.

stereoisomer One of two or more **isomers** with the same **molecular formula**, but different **configurations**; any functional groups remain the same. *E.g.* 1-bromopropane, C_3H_7Br, and 2-bromopropane, $CH_3CH(Br)CH_3$, are stereoisomers.

stereoisomerism Isomerism of compounds of the same **molecular formula** that results when the spatial arrangement of the atoms within the molecules is different. Any functional groups remain the same but may be in different positions. *See also* **geometric isomerism; optical isomerism**.

stereoregular Describing a compound that has a regular spatial arrangement of **atoms** within its **molecule**.

stereoregular rubber Member of a group of stereoregular synthetic rubbers which are produced by a solution **polymerization** process using special **catalysts**.

stereospecific Describing a **chemical reaction** in which one product is formed from each **geometric isomer** of the **reactant**.

steric effect Phenomenon in which the shape of a molecule affects its **chemical reactions**.

steric hindrance Repulsion of an approaching potential **reactant** by a sterically congested compound. *See* **steric effect**.

■ **steroid** Any of a group of naturally occurring tetracyclic organic compounds, widely found in animal tissues. Most have very important physiological activities (*e.g.* adrenal hormones, bile acids, sex hormones, sterols). [3/2/b] [3/9/b]

sterol Subgroup of **steroids** or steroid **alcohols**. They include **cholesterol**, abundant in animal tissues, which is the precursor of many other steroids.

■ **still** Apparatus for the **distillation** of liquids.

■ **stoichiometric compound** Chemical whose molecules have the component **elements** present in exact proportions as demanded by a simple **molecular formula**.

stoichiometric mixture Mixture of **reactants** that in a chemical reaction yield a **stoichiometric compound** with no excess reactant.

■ **stoichiometry** Branch of **chemistry** that deals with the relative quantities of atoms or molecules taking place in a reaction.

storage cell Alternative name for **secondary cell**.

■ **STP** or **s.t.p.** Abbreviation of **standard temperature and pressure**.

■ **straight chain Molecule** in which the **atoms** are linked in a long chain, with no **side chains** attached.

strangeness Property of **hadrons** (subatomic particles), some of which have zero strangeness, whereas others possess non-zero strangeness because they decay slower than expected and are therefore described as strange. *See also* **spin**.

■ **strong acid Acid** that is completely dissociated into its component **ions**.

■ **strong base** **Base** that is completely dissociated into its component **ions**.

strontia Alternative name for **strontium oxide**.

strontium Sr Silvery-white metallic element in Group IIA of the Periodic Table (one of the **rare earth** elements). Strontium compounds impart a bright red colour to a flame, and are used in flares and fireworks. At. no. 38; r.a.m. 87.62.

strontium nitrate $Sr(NO_3)_2$ Colourless crystalline solid, used in fireworks and flares to produce a bright red flame.

strontium oxide SrO Grey amorphous solid, which dissolves in water to form strontium hydroxide. Alternative names: strontia, strontium monoxide.

strontium unit (S.U.) Measure of the concentration of the **radio-isotope** strontium-90 in substances such as bone, milk or soil relative to their **calcium** content. *See also* **strontium**.

■ **structural formula** Shorthand description of a chemical compound that indicates the arrangement of the atoms in its molecules as well as its composition. *E.g.* the molecular formula of acetone (propanone) is C_3H_6O; its structural formula is CH_3COCH_3. *See also* **empirical formula**; **molecular formula**.

structural isomerism **Isomerism** of chemical compounds that have the same **molecular formula** but different **structural formulae**. Structural isomers have different physical and chemical properties. *E.g.* propanol (propyl alcohol), C_3H_5OH, and acetone (propanone), CH_3COCH_3, have the same molecular formulae, C_3H_6O, but different structures and are therefore structural isomers.

■ **strychnine** White crystalline insoluble **alkaloid** with a bitter taste, one of the most powerful **poisons** known. Alternative name: vauqueline.

styrene $C_6H_5CH=CH_2$ Colourless liquid **aromatic compound** that occurs in coal-tar. It polymerizes slowly on standing, rapidly when exposed to sunlight, and is used for making **polymers** (*e.g.* **polystyrene**). Alternative names: phenylethylene, vinylbenzene, ethenylbenzene.

styrene-butadiene rubber Synthetic rubber made by the **polymerization** of **butadiene** with **styrene**. It is used for making vehicle tyres. Alternative names: buna-S, SBR, GR-S.

■ **subatomic particle Particle** that is smaller than an **atom** or forms part of an atom (*e.g.* electron, neutron, proton). Sometimes also called an elementary particle. [8/8/a]

subcritical Describing a **chain reaction** in a **nuclear reactor** that is not self-sustaining.

■ **sublimate** Solid formed by the process of **sublimation**.

■ **sublimation** Direct conversion of a solid substance to its vapour state on heating without melting taking place (*e.g.* solid carbon dioxide (dry ice), iodine). The vapour condenses to give a **sublimate**. The process is used to purify various substances.

sub-shell Subdivision of an electron **shell**.

substituent Atom or group that replaces another atom or group in a molecule of a compound.

■ **substitution** Direct replacement of an **atom** or group in a molecule of a compound by some other atom or group.

■ **substitution product** Product formed from **substitution**.

■ **substitution reaction** Chemical reaction that involves the direct replacement of an **atom** or group in a **molecule** of a compound by some other atom or group. *E.g.* the reaction in which an atom of chlorine replaces an atom of hydrogen in a molecule of benzene (C_6H_6) to form a molecule of chlorobenzene (C_6H_5Cl). Alternative name: displacement reaction. *See also* **addition reaction**.

substrate *1*. Molecule on which an **enzyme** exerts its catalytic action. *2*. Pure crystal of a **semiconductor** used for making integrated circuits.

succinic acid $(CH_2COOH)_2$ White crystalline **dicarboxylic acid**, used in the manufacture of dyes and in organic synthesis. Alternative names: butanedioic acid, ethylenedicarboxylic acid.

sucrase Enzyme that breaks down **sucrose** into simpler sugars. Alternative name: invertase.

■ **sucrose** $C_{12}H_{22}O_{11}$ White **optically active** soluble crystalline disaccharide which is obtained from sugar cane and sugar-beet; ordinary sugar, used to sweeten food. It is hydrolysed to **fructose** and **glucose**. Alternative names: cane-sugar, beet-sugar, sugar. [3/6/b]

■ **sugar** *1*. Crystalline soluble **carbohydrate** with a sweet taste; usually a **monosaccharide** or **disaccharide**. *2*. Common name for **sucrose**.

sulphate Ester or **salt** of **sulphuric acid**.

sulphation Conversion of a substance into a **sulphate**.

sulphide Binary compound containing **sulphur**; a salt of **hydrogen sulphide**. Organic sulphides are called **thioethers**.

sulphite Ester or **salt** of **sulphurous acid**.

2-sulphobenzimide Alternative name for **saccharin**.

sulphonamide Member of a class of synthetic anti-bacterial drugs which act by **enzyme** inhibition. They are **amides** derived from **sulphonic acids**.

sulphonate Ester or **salt** of a **sulphonic acid**.

sulphonation Substitution reaction that involves the replacement of a **hydrogen** atom by the **sulphonic acid** group.

sulphonic acid Acid that contains the group $-SO_3H$. Organic sulphonic acids are used in the manufacture of dyes, detergents and drugs.

sulphonium compound Compound of the general formula R_3SX, where R is an organic **radical** and X is an electronegative element or radical.

sulphoxide Compound of general formula RSOR′, where R and R′ are organic **radicals**.

■ **sulphur** S Yellow nonmetallic solid element in Group VIB of the Periodic Table, which forms several **allotropes** including alpha- (rhombic) sulphur and beta- (monoclinic) sulphur. It occurs as the free element in volcanic regions and as underground deposits (extracted by the **Frasch process**), and as **sulphates** and **sulphides**, which include important minerals (*e.g.* galena, PbS, and pyrites, FeS_2). Chemically it behaves like oxygen, and can replace it in organic compounds (*e.g.* **thioethers** and **thiols**). Sulphur is used to make **sulphuric acid**, matches, gunpowder, drugs, fungicides and dyes, and in the **vulcanization** of rubber. At. no. 16; r.a.m. 32.06.

sulphur dichloride oxide Alternative name for **thionyl chloride**.

■ **sulphur dioxide** SO_2 Colourless poisonous gas with a strong pungent odour, made by burning sulphur, roasting **sulphide** ores or by the action of acids on **sulphites**. It is used to make **sulphuric acid** and, in aqueous solution, as a bleach (*e.g.* for straw and paper). Alternative name: sulphur(IV) oxide.

sulphur dye Dye made by heating certain organic compounds with **sulphur** or alkali polysulphides, used for dyeing industrial fabrics.

■ **sulphuric acid** H_2SO_4 Corrosive colourless oily liquid **acid**, made mainly from **sulphur dioxide** by the **contact process**. It is a **desiccant**, and when hot a powerful **oxidizing agent**. It is produced in large quantities and used in the manufacture of other acids, fertilizers, explosives, accumulators, petrochemicals, etc. Its salts are **sulphates**. Alternative names: vitriol, oil of vitriol, hydrogen sulphate.

■ **sulphurous acid** H_2SO_3 Colourless aqueous solution of **sulphur dioxide**, used as a bleach and a **reducing agent**. Its salts are **sulphites**.

■ **sulphur(IV) oxide** Alternative name for **sulphur dioxide**.

■ **sulphur(VI) oxide** Alternative name for **sulphur trioxide**.

■ **sulphur trioxide** SO_3 Volatile white solid made by the catalytic oxidation of **sulphur dioxide**, usually stored in sealed tubes. It reacts with water to form **sulphuric acid**. Alternative name: sulphur(VI) oxide.

supercritical Describing a **chain reaction** of a **nuclear reactor** which is self-sustaining.

supernatant liquid Clear liquid that lies above a sediment or **precipitate**.

superoxide *1.* Compound that yields the **free radical** O_2^- which is highly toxic to living cells. *2.* **Oxide** that yields both **hydrogen peroxide** and **oxygen** on treatment with an **acid**. *See also* **peroxide**.

■ **superphosphate** Important **fertilizer** made by reacting calcium phosphate (from slag or bone ash) with **sulphuric acid**.

■ **supersaturated solution** Unstable **solution** that contains more

solute than a **saturated solution** would contain at the same temperature. It easily changes to a saturated solution when the excess solute is made to crystallize.

■ **surface active agent** Alternative name for **surfactant**.

■ **surfactant** Substance that reduces the surface tension of a liquid, used in **detergents**, wetting agents and foaming agents. Alternative name: surface active agent.

■ **suspension** Mixture of insoluble small solid particles and a liquid in which the insoluble substance stays evenly distributed throughout the liquid. Alternative name: suspensoid.

■ **symbol, chemical** Letter or letters that represent the name of a **chemical element**, or one atom of it in a chemical **formula**; *e.g.* C is carbon, Ca is calcium and Co is cobalt. The symbol may be based on the element's Latin, not English, name; *e.g.* Au for gold (Latin *aurum*), Fe for iron (*ferrum*).

■ **synthesis** Formation of a chemical compound by combining elements or simpler compounds; the building of compounds through a planned series of steps.

■ **synthetic** Formed by artificial means; describing a chemical compound that has been produced by synthesis. Alternative names: artificial, man-made.

synthetic gas Mixture of **hydrogen** and **carbon monoxide** made by **reforming** natural gas with steam.

T

tachyon Theoretical subatomic particle that travels faster than light.

■ **talc** $3MgO.4SiO_2.H_2O$ Soft white or grey-green mineral. Its purified form is a white powder, used in talcum powder, in medicine and in ceramic materials. Alternative names: French chalk, magnesium silicate monohydrate.

tall oil Complex mixture of mainly **resin** acids and **fatty acids**, a by-product of wood-pulp manufacture, used in **soaps**, paints and varnishes.

tannic acid White amorphous solid organic acid, a member of the class of compounds called **tannins**. It is used in **tanning** and for making inks and dyes. Alternative name: tannin.

tannin Yellow substance that is a member of a class of organic compounds, of vegetable origin (*e.g.* in tree bark, oak galls and tea), that are derivatives of **polyhydric benzoic acids**. They are used in tanning hides to make leather. *See also* **tannic acid**.

■ **tanning** Process of converting hides and skin into leather by the action of substances such as **tannic acid** or **tannins**.

tantalum Ta Hard blue-grey metallic element in Group VA of the Periodic Table (a **transition element**), used in electronic and chemical equipment, and for making surgical instruments. At. no. 73; r.a.m. 180.948.

■ **tar** Dark viscous liquid obtained by the **destructive distillation** of **coal** in the absence of air. It is a complex mixture of organic compounds, mainly **hydrocarbons** and **phenols**.

Alternative name: coal-tar. Similar substances obtained from **petroleum** are also called tars. *See also* **bitumen**.

tartar emetic Potassium antimonyl tartrate, $K.SbO.C_4H_4O_6.\frac{1}{2}H_2O$, a poisonous compound used as an insecticide and as mordant in dyeing.

tartaric acid HOOC.CH(OH)CH(OH).COOH White crystalline hydroxycarboxylic acid that occurs in grapes and other fruits. It is used in dyeing and printing. Its salts, the **tartrates**, are frequently used in medicine. Alternative names: 2,3-dihydroxybutanedioic acid, dihydroxysuccinic acid.

tartrate Ester or **salt** of **tartaric acid**.

tau-particle Subatomic particle, a **lepton** and its associated anti-lepton that have unusually high mass.

tautomerism Equilibrium between two organic **isomers**. It usually involves a shift in the point of attachment of a mobile hydrogen atom and a shift in the position of a **double bond** in a molecule. Alternative name: dynamic isomerism. *See* **keto-enol tautomerism**.

■ **tear gas** Volatile substance, usually a **halogen**-containing organic compound, that causes irritation of the eyes and is used in crowd control. Alternative name: lachrymator.

technetium Tc Artificial radioactive metallic element in Group VIIA of the Periodic Table (a **transition element**), which occurs among the fission products of uranium. It has several isotopes, with half-lives of up to 2.12×10^5 years. At. no. 43; r.a.m. 99 (most stable isotope).

■ **Teflon** Trade name for **polytetrafluoroethene** (PTFE).

telluride Binary compound that contains **tellurium**.

tellurium Te Silvery-white semi-metallic element in Group

VIB of the Periodic Table, obtained as a by-product of the extraction of gold, silver and copper. It is used as a catalyst and to add hardness to alloys of lead or steel. At. no. 52; r.a.m. 127.60.

■ **tempering** Method of changing the physical properties of a metal or alloy by heating it for a time and then cooling it gradually or quickly.

■ **temporary hardness** *See* **hardness of water**. [5/6/a]

terbium Tb Silvery-grey metallic element in Group IIIA of the Periodic Table (one of the **lanthanides**), used in making **semiconductors** and **phosphors**. At. no. 65; r.a.m. 158.92.

terephthalic acid $C_6H_4(COOH)_2$ White crystalline **aromatic compound,** used in the manufacture of **plasticizers** and **polyester** fibres and films (*e.g.* Dacron, Terylene). Alternative name: benzene-1,4-dicarboxylic acid. *See also* **phthalic acid**.

terpene Member of a class of **unsaturated hydrocarbons** found in **essential oils**, with a structure based on the **isoprene** unit.

tervalent Alternative name for **trivalent**.

■ **Terylene** Trade name for a **polyester** produced by the **condensation** of **terephthalic acid** with ethane-1,2-diol (ethylene glycol). It is used for making textiles.

■ **tetrachloroethene** $CCl_2=CCl_2$ Liquid halogenoalkene, used as a solvent (especially as a de-greasing agent). Alternative names: ethylene tetrachloride, tetrachloroethylene.

■ **tetrachloroethylene** Alternative name for **tetrachloroethene**.

■ **tetrachloromethane** CCl_4 Colourless liquid **halogenoalkane,** used as a solvent, cleaning agent and in fire extinguishers. Alternative name: carbon tetrachloride.

■ **tetraethyllead** $Pb(C_2H_5)_4$ Colourless liquid **organometallic compound,** used as an anti-**knock** agent in petrol. Alternative name: tetraethyl plumbane.

tetrafluoroethene $CF_2{=}CF_2$ Gaseous **fluorocarbon** from which **polytetrafluoroethene** (PTFE, Fluon) is made.

■ **tetrahedral compound** Compound that has a central **atom** joined to other atoms or groups at the four corners of a tetrahedron. It is the configuration of carbon in many **saturated compounds** (*e.g.* in methane, CH_4).

■ **tetrahydrate** Chemical containing four molecules of **water of crystallization.**

■ **tetravalent** Having a **valence** of four. Alternative name: quadrivalent.

thallium Tl Silver-grey metallic element in Group IIIB of the Periodic Table, used in electronic equipment and to make pesticides. At. no. 81; r.a.m. 204.39.

thermal analysis *1.* Method of chemical analysis that depends on a change of weight of a compound as a function of temperature. Alternative name: thermographic analysis. *2.* Method of analysing metals and alloys by studying their heating and cooling curves.

thermal cross-section Probability of a **nucleus** interacting with a **neutron** of thermal energy (*i.e.* a slow neutron).

thermal diffusion Process of forming a concentration gradient in a **fluid** mixture by the application of a temperature gradient. Alternative name: Soret effect.

■ **thermal dissociation** Temporary reversible decomposition of a chemical compound (into component molecules) resulting from the application of heat.

thermal equilibrium State in which there is no net heat flow in a system.

thermalization Slowing-down of fast **neutrons** to thermal energies (*i.e.* converting them to slow neutrons, capable of initiating **nuclear fission**), achieved by a **moderator** in a **nuclear reactor**.

thermal neutron Neutron that has an energy of about 0.025 electron volts (eV), and which can be captured by an atomic **nucleus** (particularly to initiate **nuclear fission**). Alternative name: slow neutron. *See also* **fast neutron**.

thermal reactor Nuclear reactor that has a **moderator,** and in which a **chain reaction** is sustained by **thermal neutrons**.

thermite Mixture of powdered aluminium and iron(III) oxide (ferric oxide) which produces a great amount of heat when ignited, because the reaction between the two substances is highly **exothermic**. It is used in incendiary bombs and for local welding of steel (*e.g.* to mend broken tram-lines). Alternative name: Thermit.

■ **thermochemistry** Branch of **chemistry** that deals with the study of heat changes in relation to **chemical reactions**. The measurement of **heats of reaction**, heat capacities and **bond energies** falls within its scope.

thermographic analysis Alternative name for **thermal analysis**.

■ **thermonuclear reaction** Process that releases energy by the fusion of atomic nuclei (*e.g.* of hydrogen nuclei in the **hydrogen bomb**). Alternative name: fusion reaction. *See also* **nuclear fusion.**

■ **thermoplastic** Describing a **polymer** (plastic) that softens on heating and hardens on cooling without changing its properties (*i.e.* it softens again on reheating). The polymer itself may be described as a thermoplastic. *See also* **thermosetting**. [7/9/b]

■ **thermosetting** Describing a **polymer** (plastic) that becomes rigid on heating because of the formation of extra chemical bonds, and therefore has new properties. It does not soften again on reheating – heat will decompose it. The polymer may be described as a thermosetter. *See also* **thermoplastic**. [7/9/b]

thiamine White water-soluble crystalline B **vitamin**, found in cereals and yeast. Deficiency of thiamine causes the disorder beri-beri in human beings. Alternative names: thiamin, aneurin, vitamin B_1.

thiazine Member of a group of **heterocyclic compounds** that contain **sulphur** and **nitrogen** (in addition to four carbons) in a six-membered ring.

thin layer chromatography (TLC) Method for separating substances by allowing a **solution** containing them to rise up a thin plate of solid **adsorbent** material, with the different substances moving at different rates. Once separated, they can be identified.

■ **thio-** Prefix denoting the presence of **sulphur** in a compound.

thiocarbamide Alternative name for **thiourea**.

thioether Member of a class of organic compounds, general formula RSR′ where R and R′ are **alkyl** or **aryl groups**, analogous to **ethers**, in which **sulphur** takes the place of **oxygen**. They are named as alkyl or aryl sulphides; *e.g.* ethyl methyl sulphide, $C_2H_5SCH_3$.

thiol Member of a class of unpleasant-smelling organic compounds, general formula RSH, where R is an **alkyl** or **aryl**

group, analogous to **alcohols**, in which **sulphur** takes the place of **oxygen** in the **hydroxyl group**; *e.g.* methane thiol, CH_3SH. The group –SH is called the thiol or mercapto group. Alternative name: mercaptans.

thionyl chloride $SOCl_2$ Colourless fuming pungent liquid, used in organic synthesis (to convert an –OH group to –Cl). Alternative name: sulphur dichloride oxide.

thiophene C_4H_4S Colourless liquid that occurs as an impurity in commercial **benzene**. A **heterocyclic** compound, with four carbon atoms and one sulphur atom in a five-membered ring, it is used in organic synthesis and as a solvent.

thiosulphate Ester or **salt** of **thiosulphuric acid**. The alkali metal salts, *e.g.* sodium thiosulphate, $Na_2S_2O_3$, and ammonium thiosulphate, $(NH_4)_2S_2O_3$, are used in photography as fixing agents (where they dissolve unexposed silver halides from the emulsion on developed film or paper).

thiosulphuric acid $H_2S_2O_3$ Unstable **acid** which readily decomposes to **sulphurous acid** and **sulphur**. Its salts (**thiosulphates**) are used in photography.

thiourea NH_2CSNH_2 Colourless crystalline organic compound, the **sulphur** analogue of **urea**. Its conversion to its **isomer** ammonium thiocyanate, NH_4CNS, on heating was the first demonstration of an organic compound being changed directly into an inorganic one. It is used in medicine and as a photographic sensitizer. Alternative name: thiocarbamide.

thoria Alternative name for **thorium dioxide**.

thorium Th Silvery-white radioactive element in Group IIIA of the Periodic Table (one of the **actinides**). It has several isotopes, with half-lives of up to 1.39×10^{10} years. Thorium-232 captures slow, or thermal, neutrons and is used to 'breed' the fissile uranium-233. Its refractory oxide (thoria, ThO_2) is used in gas mantles. At.no. 90; r.a.m. 232.0381.

thorium dioxide ThO_2 White insoluble powder, used as a refractory and in non-silica optical glass. Alternative name: thoria.

threonine **Amino acid** that is essential in the diet of animals. Alternative name: 2-amino-3-hydroxybutanoic acid.

thulium Tm Silvery-white metallic element in Group IIA of the Periodic Table (one of the **lanthanides**). Its radioactive isotopes emit gamma-rays and X-rays and are used in portable radiography equipment. At. no. 69; r.a.m. 168.934.

thymine $C_5H_6N_2O_2$ Colourless crystalline **heterocyclic** compound, one of the **pyrimidines**. Alternative names: 5-methyluracil, 5-methyl-2,4-dioxopyrimidine.

thymol $C_{10}H_{14}O$ Colourless crystalline organic compound found in the **essential oils** of thyme and mint. It is used in antiseptic mouth washes. Alternative name: 2-hydroxy-*p*-cymene, 2-hydroxy-1-isopropyl-4-methylbenzene.

■ **tin** Sn Soft silvery-white metallic element in Group IVB of the Periodic Table, which forms three **allotropes**. It occurs mainly as tin(IV) oxide, SnO_2, in ores such as cassiterite (tinstone). It is used mainly as a protective coating for steel (tin plate) and in making alloys with lead (solder, type metal, pewter). Its compounds are used as catalysts, fungicides and mordants. At. no. 50; r.a.m. 118.69.

tin(II) Alternative name for stannous.

tin(IV) Alternative name for stannic.

tincal Naturally occurring crude **borax**.

tin(II) chloride $SnCl_2$ White soluble solid. A **reducing agent**, it is used as a **catalyst** in organic reactions and as an anti-sludge agent for oils. Alternative names: stannous chloride, tin salt.

tin(IV) chloride $SnCl_4$ Colourless fuming liquid, used to coat **glass** with **tin(IV) oxide** (to make it conductive), as a mordant and in the preparation of inorganic tin and organotin compounds. Alternative name: stannic chloride.

tin dioxide Alternative name for **tin(IV) oxide**.

tin disulphide Alternative name for **tin(IV) sulphide**.

tin(IV) hydride SnH_4 Unstable gas, used as a **reducing agent** in organic chemistry. Alternative name: stannane.

tin hydroxide oxide Alternative name for **stannic acid**.

tin(IV) oxide SnO_2 White crystalline solid, used as a **pigment** and as a refractory material. Alternative name: tin dioxide.

tin salt Alternative name for **tin(II) chloride**.

tin(IV) sulphide SnS_2 Yellow insoluble solid, used as a **pigment**. Alternative name: tin disulphide.

titania Alternative name for **titanium(IV) oxide**.

titanic chloride Alternative name for **titanium(IV) chloride**.

■ **titanium** Ti Silvery-white metallic element in Group IVA of the Periodic Table (a **transition element**). Its corrosion-resistant lightweight alloys are employed in the aerospace industry. Naturally occurring crystalline forms of titanium(IV) oxide (titania, TiO_2) constitute the semi-precious gemstone rutile. The powdered oxide is used as a white pigment and a dielectric in capacitors. At. no. 22; r.a.m. 47.9.

titanium(IV) chloride $TiCl_4$ Colourless fuming liquid, a source of pure **titanium(IV) oxide**. Alternative names: titanium tetrachloride, titanic chloride.

titanium dioxide Alternative name for **titanium(IV) oxide**.

titanium(IV) oxide TiO_2 White inert solid that occurs in three crystalline forms: rutile, brookite and anatase. It is used as a white **pigment** and as a component in various dielectrics for electrical capacitors. Alternative names: titanium dioxide, titania.

titanium tetrachloride Alternative name for **titanium(IV) chloride**.

titrant Chemical solution of known concentration, *i.e.* a **standard solution**, which is added during the course of a **titration**.

■ **titration** Technique in **volumetric analysis** in which one chemical solution of known concentration is added (using a **burette**) to a known volume of another chemical solution of unknown concentration (measured by a **pipette**), and the **chemical reaction** followed by observing changes in colour, **pH**, etc. An **indicator** may be added to indicate the end-point of the reaction, which allows the unknown concentration to be determined.

titre Concentration of a solution as determined by **titration**.

■ **TNT** Abbreviation of **trinitrotoluene**.

tocopherol **Vitamin** isolated from plants that increases fertility in rats. Deficiency of it causes wasting of muscles in animals. It has been found to have **antioxidant** activity, and it is important in maintaining membranes. Alternative name: vitamin E.

■ **Tollens' reagent** Ammoniacal solution of **silver oxide** used as

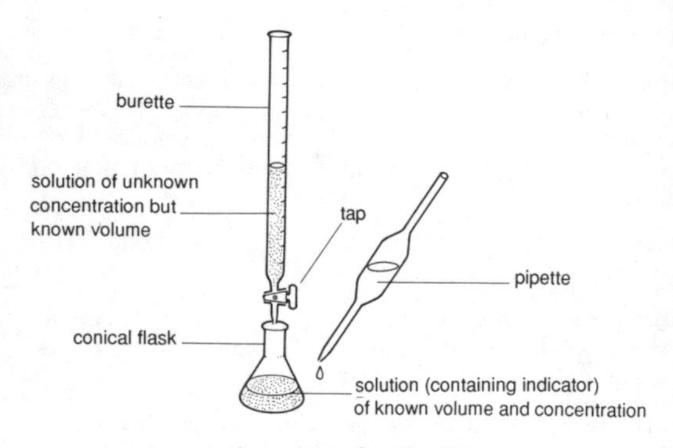

Apparatus for titration

a test for **aldehydes,** which reduce it to deposit a mirror of silver. It was named after the German chemist Bernhard Tollens (1841–1918).

■ **toluene** $C_6H_5CH_3$ Colourless aromatic organic liquid that occurs in coal-tar, used as an industrial solvent and starting point for making explosives. Alternative name: methylbenzene.

toluidine $CH_3C_6H_4NH_2$ One of three isomeric aromatic **amines**, used in the manufacture of dyes and drugs. The ortho- and meta- forms are colourless liquids; the para-isomer is a colourless crystalline solid. Alternative names: aminotoluene, methylaniline.

topaz $Al_2SiO_4(OH,F)_2$. Usually pale yellow semi-precious gemstone consisting of the mineral aluminium silicate with fluoride and hydroxyl ions.

toxin *1.* Poison produced by bacteria or other biological sources. *2.* Any harmful substance.

toxoid Bacterial **toxin** that has been chemically treated to make it non-poisonous, used as a vaccine.

trace element Element essential to metabolism, but necessary only in very small quantities (*e.g.* copper and cobalt in animals, molybdenum in plants). Such elements are usually toxic if large quantities are ingested.

transamination Removal and transference of an **amino group** from one compound (usually an **amino acid**) to another.

transference number Alternative name for **transport number**.

transformation, nuclear Transmutation of one atomic **nucleus** into another by means of a nuclear reaction.

transformation constant Alternative name for **disintegration constant**.

transition element Member of a large group of elements that have partly filled inner electron shells, which gives them their distinctive physical and chemical properties (particularly variable valence and the tendency to form coloured compounds). They occupy Groups IIIA, IVA, VA, VIA, VIIA, VIII, IB and IIB of the Periodic Table. Many of these elements and their compounds are used as **catalysts**.

transition temperature Temperature above and below which different **allotropes** are stable.

transport number In **electrolysis**, fraction of the total current carried by a particular **ion** in the **electrolyte**. Alternative name: transference number.

transuranic element Element with a higher atomic number than **uranium** (92). Transuranic elements are artificially made and **radioactive**. Alternative names: transuranium element, uranide. *See also* **post-actinide**.

■ **tri-** Prefix meaning three.

triaminotriazine Alternative name for **melamine**.

■ **triatomic molecule** Molecule of an element that consists of three **atoms**, *e.g.* **ozone**, O_3.

triazine $C_3H_3N_3$ One of a group of isomeric **heterocyclic** organic compounds with three nitrogen atoms and three carbon atoms in the ring. Triazine derivatives are used as dyes and herbicides.

triazole $C_2H_3N_3$ One of a group of isomeric **heterocyclic** organic compounds with three nitrogen atoms and two carbon atoms in the ring.

■ **tribasic acid Acid** that has three replaceable hydrogen atoms in its molecules, *e.g.* orthophosphoric acid, H_3PO_4.

tribromomethane $CHBr_3$ Colourless liquid **haloform**, used for the separation of minerals and in organic synthesis. Alternative name: bromoform.

trichloroacetaldehyde Alternative name for **trichloroethanal**.

trichloroethanal CCl_3CHO Pungent colourless oily liquid **aldehyde**, which forms a solid hydrate (**trichloroethanediol**). Alternative names: chloral, trichloroacetaldehyde.

1,1,1-trichloroethane Cl_3CCH_3 Colourless aromatic liquid, used as an industrial solvent and de-greasing agent. Alternative name: methyl chloroform.

trichloroethanediol $Cl_3CCH(OH)_2$ White crystalline organic compound, used as a sedative. Alternative name: chloral hydrate.

trichloromethane $CHCl_3$ Colourless volatile liquid **haloform**, used as an anaesthetic and as a solvent. Alternative name: chloroform.

triglyceride Ester of **glycerol**, in which all three **hydroxyl groups** have been substituted by ester groupings from **fatty acids**. Many **fats** are triglycerides.

triiodomethane CHI_3 Yellow crystalline solid **haloform**, used as an antiseptic. Alternative name: iodoform.

trimer Chemical formed by the combination of three similar (monomer) molecules; *e.g.* three molecules of **acetaldehyde** (ethanal), CH_3CHO, combine to form a single cyclic molecule of **paraldehyde** (ethanal trimer), $C_6H_{12}O_3$.

trimethylaluminium $Al(CH_3)_3$ Colourless reactive liquid, used in organic synthesis. Alternative name: aluminium trimethyl.

trinitroglycerine Alternative name for **nitroglycerine**, or glyceryl trinitrate.

trinitrophenol Alternative name for **picric acid**.

trinitrotoluene $CH_3C_6H_2(NO_2)_3$ Yellow crystalline aromatic organic compound, used as a high explosive. Alternative names: methyl-2,4,6-trinitrobenzene, TNT.

triple bond Covalent bond formed by the sharing of three pairs of **electrons** between two **atoms**.

triple point Temperature and pressure at which the three phases (solid, liquid and vapour) of a substance are in equilibrium. The triple point of water occurs at 273.16 K and 611.2 Pa. *See also* **Kelvin temperature**.

trisaccharide Carbohydrate consisting of three joined **monosaccharides**.

■ **tritium** T Radioactive **isotope** of **hydrogen**; it has two neutrons and one proton in its nucleus. R.a.m. 3.016.

triton Atomic nucleus of **tritium**, consisting of two neutrons and one proton.

■ **trivalent** Having a **valence** of three. Alternative name: tervalent.

■ **trypsin** Digestive **enzyme** secreted into the small intestine, where it catalyses the **hydrolysis** of **polypeptide** chains (of proteins) at specific sites.

tryptophan Essential amino acid that contains an aromatic group, needed in animals for proper growth and development.

■ **tungsten** W Steel-grey metallic element in Group VIA of the Periodic Table (a **transition element**) of high melting point and great hardness, used for making electric filaments and special steels for turbine blades and cutting tools. **Tungsten carbides** are also extremely hard. At. no. 74; r.a.m. 183.85. Alternative name: wolfram.

■ **tungsten carbide** WC and WC_2. Hard refractory substances, used for the tips of cutting tools and as abrasives.

■ **turpentine** Yellow sticky natural **resin** obtained from various coniferous trees. Alternative name: pine-cone oil.

turpentine oil Colourless volatile liquid; an **essential oil** obtained from the distillation of **turpentine**. It is used as a solvent for varnishes and polishes, and in medicine. Alternative name: oil of turpentine.

turquoise Blue mineral; naturally occurring copper aluminium phosphate. It is used as a gemstone. Alternative names: callaite, callanite.

tyrosine White crystalline organic compound, a naturally occurring **essential amino acid** found in most **proteins**, and a precursor in the body of various **hormones**.

U

uniaxial crystal Doubly refracting crystal in which there is only one direction of single refraction.

unimolecular reaction Chemical reaction that involves only one type of **molecule** as the **reactant**; *e.g.* the decomposition of mercury(II) oxide, HgO, on strong heating to give mercury and oxygen.

unit cell Smallest group of **atoms, ions** or **molecules** whose three-dimensional repetition at regular intervals produces a **crystal lattice**.

univalent Alternative name for **monovalent**.

■ **universal indicator** Mixture of chemical **indicators** that gives a definite colour change for various values of **pH**. [6/5/b]

■ **unsaturated** *1.* Describing an organic compound with doubly or triply bonded carbon atoms; *e.g.* ethene (ethylene), $C_2=H_4$, and ethyne (acetylene), $C_2\equiv H_2$. *See also* **saturated compound**. *2.* Describing a solution that can dissolve more **solute** before reaching **saturation**.

urania Alternative name for **uranium(IV) oxide**.

uraninite Dark-coloured, dense and slightly greasy mineral, one of the major ores of **uranium**. Alternative name: pitchblende.

■ **uranium** U Radioactive grey metallic element in Group IIIA of the Periodic Table (one of the **actinides**), obtained mainly from its ore uraninite (which contains uranium(IV) oxide, UO_2). It has three natural and several artificial **isotopes** with half-lives of up to 4.5×10^9 years. Uranium-235 undergoes

nuclear fission and is used in nuclear weapons and reactors; uranium-238 can be converted into the fissile plutonium-239 in a **breeder reactor**. At. no. 92; r.a.m. 238.03. [8/7/d]

uranium dioxide Alternative name for **uranium(IV) oxide**.

uranium-lead dating Dating method used in calculating the geologic age of minerals based on the **radioactive decay** of **uranium**-235 to lead-207 and **uranium**-238 to lead-206.

uranium(IV) oxide UO_2 Highly toxic **radioactive** spontaneously inflammable black crystalline solid, used in photographic chemicals, ceramics, pigments and packing of nuclear fuel rods. Alternative names: urania, uranium dioxide.

uranium(VI) oxide UO_3 Highly toxic **radioactive** orange powder, used in uranium refining, as a pigment and in ceramics. Alternative names: uranium trioxide, orange oxide.

uranium trioxide Alternative name for **uranium(VI) oxide**.

uranyl Radical UO_2^{2+}, *e.g.* as in uranyl nitrate $UO_2(NO_3)_2$.

■ **urea** H_2NCONH_2 White crystalline organic compound, found in the urine of mammals as the natural end-product of the metabolism of **proteins**. It is also manufactured commercially from **carbon dioxide** and **ammonia** under high pressure. It is used in plastics, adhesives, fertilizers and animal-feed additives. Alternative name: carbamide.

■ **urea-formaldehyde resin** Colourless, non-inflammable, weather-resistant **thermosetting** plastic, manufactured by the reaction of **urea** with **formaldehyde** (methanal) or its **polymers**. Alternative name: urea resin.

urease Enzyme that occurs in plants (*e.g.* soya beans) and acts as a **catalyst** for the **hydrolysis** of **urea** to **ammonia** and **carbon dioxide**.

urethane $CO(NH_2)OC_2H_5$. Highly toxic inflammable organic compound, used in veterinary medicine, biochemical research and as a solvent and chemical intermediate. Alternative names: ethyl carbamate, ethyl urethane.

urethane resin *See* **polyurethane**.

uric acid $C_5H_4N_4O_3$ White crystalline organic acid of the **purine** group, the end-product of the metabolism of **amino acids** in reptiles and birds. In human beings uric acid deposition in the joints is the principal cause of gout.

V

■ **vacuum distillation Distillation** under reduced pressure, which helps to lower the boiling point and hence reduce the risk of **thermal dissociation**. Alternative name: reduced-pressure distillation.

■ **valence** Positive number that characterizes the combining power of an **atom** of a given **element** to the number of hydrogen atoms or their equivalent (in a chemical reaction). For an ion, the valence equals the charge on the ion. Alternative name: valency.

valence band *1.* Highest **energy level** in an **insulator** or **semiconductor** that can be filled with **electrons**. *2.* Region of electronic energy level that binds **atoms** of a **crystal** together.

■ **valence bond Chemical bond** formed by the interaction of **valence electrons** between two or more **atoms**.

■ **valence electron Electron** in an outer **shell** of an **atom** which participates in bonding to other atoms to form **molecules**.

valine $C_5H_{11}NO_2$ One of the **essential amino acids** required for normal growth in animals. Alternative names: 2-aminoisovaleric acid, 2-amino-3-methylbutyric acid.

■ **vanadium** V Silvery-grey metallic element in Group VA of the Periodic Table (a **transition element**), used to make special steels. Vanadium(V) oxide, V_2O_5, is used as an industrial **catalyst** and in ceramics. At. no. 23; r.a.m. 50.94.

van der Waals' equation (of state) Equation of state that takes into account both the volume of the gas molecules and the attractive forces between them. It may be represented as

$(P + a/V^2)(V - b) = RT$, where V is the volume per mole, P the pressure, T the absolute temperature, R the gas constant, and a and b are constants for a given gas, evaluated by fitting the equation to experimental *PVT* measurements at moderate densities. It was named after the Dutch physicist Johannes van der Waals (1837–1923).

van der Waals' force Weak attractive force induced by interaction of dipole moments between atoms or non-polar molecules. It is represented by the coefficient a in **van der Waals' equation**.

van't Hoff's law **Osmotic pressure** of a solution is equal to the pressure that would be exerted by the **solute** if it were in the gaseous phase and occupying the same volume as the solution at the same temperature. It was named after the Dutch chemist Jacobus van't Hoff (1852–1911).

■ **vapour** A **gas** when its temperature is below the critical value; a vapour can thus be condensed to a liquid by pressure alone.

■ **vapour density** Density of a gas relative to a reference gas, such as hydrogen, equal to the mass of a volume of gas divided by the mass of an equal volume of hydrogen at the same temperature and pressure. It is also equal to half the **relative molecular mass**.

vapour pressure **Pressure** under which a liquid and its vapour coexist at equilibrium. Alternative name: saturation vapour pressure.

vauqueline Alternative name for **strychnine**.

velocity constant Alternative name for **rate constant**.

■ **verdigris** Green copper(II) carbonate, $CuCO_3.Cu(OH)_2$, formed by corrosion of metallic copper or its alloys. The term is also used for the similar basic copper(II) acetate, used as a pigment, fungicide and mordant in dyeing.

vesicant Blister-causing agent (*e.g.* mustard gas).

■ **vinegar** Dilute solution (about 4 per cent by volume) of **acetic acid** (ethanoic acid). Natural vinegar is made by bacterial fermentation of cider or wine.

vinyl acetate $CH_2=CHOOCCH_3$ **Monomer** from which the plastic and adhesive polyvinyl acetate (PVA) is made. Alternative name: ethenyl ethanoate.

vinyl benzene Alternative name for **styrene**.

vinyl chloride $CH_2=CHCl$ **Monomer** from which the plastic (polymer) polyvinyl chloride (PVC) is made. Alternative name: chloroethene.

vinyl cyanide Alternative name for **acrylonitrile**.

vinyl group Double-bonded organic group $CH_2=CH-$.

virgin neutron Any **neutron** from any source before collision.

■ **viscose rayon** *See* **rayon**.

■ **vitamin** Any of several organic compounds which in small quantities are essential to the proper growth and regulation of metabolic processes, *e.g.* energy transformation in animal organisms. There are two major groups, water-soluble (*e.g.* vitamins C, B) and fat-soluble (*e.g.* A, D, E, K), which are present in foodstuffs and must be taken as part of a balanced diet. [3/6/c]

vitriol Alternative name for **sulphuric acid**.

■ **volatile** Describing any substance that is readily changed to a **vapour** and hence lost through **evaporation**. Volatile liquids have low boiling points.

■ **voltaic cell** Any device that produces an electromotive force (e.m.f.) by the conversion of chemical energy to electrical energy, *e.g.* a battery or accumulator. Alternative name: galvanic cell.

■ **volumetric analysis** Method of chemical analysis that relies on the accurate measurement of the reacting volumes of substances in solution (*e.g.* by carrying out a **titration**).

■ **vulcanization** Method of hardening natural or artificial **rubber** by heating it with sulphur or sulphur compounds.

W

Wacker process Commercial organic synthesis that uses oxygen to oxidize ethylene (ethene) to acetaldehyde (ethanal) in the presence of palladium chloride and copper(II) chloride **catalysts**. It was named after the German industrial chemist Alexander von Wacker (1846–1922).

■ **washing soda** Alternative name for hydrated **sodium carbonate**.

■ **water** H_2O Colourless liquid, one of the oxides of hydrogen (the other is **hydrogen peroxide**, H_2O_2) and the commonest substance on Earth. It can be made by burning hydrogen or fuels containing it in air or oxygen, or by the action of an **acid** on an **alkali** or **alcohol**. It is a good (polar) solvent, particularly for ionic compounds, with which it may form solid **hydrates**. It can be decomposed by the action of certain reactive metals (*e.g.* the **alkali metals**) or by **electrolysis**. Water is essential for life and forms the major part of most body fluids (*e.g.* blood, lymph). It freezes at 0°C and boils at 100°C (at normal atmospheric pressure), and has its maximum density at 3.98°C. *See also* **hardness of water**.

■ **water cycle** Continuous movement of water between the atmosphere and the land and oceans. Water that falls as precipitation (*e.g.* rain, snow, hail) passes into the ground or runs off to form springs and rivers, which ultimately flow into the oceans. Water evaporates from the oceans as vapour, forms clouds in the atmosphere, and falls again as precipitation. Some water is also returned to the atmosphere through transpiration and respiration after having been taken in by plants and animals. Alternative name: hydrological cycle.

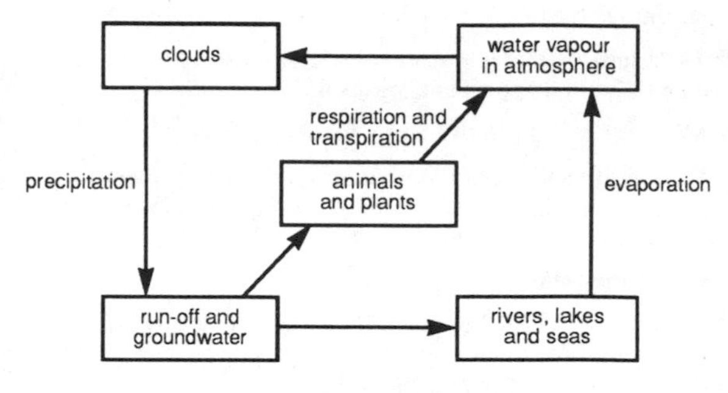

Water cycle

■ **water glass** Alternative name for **sodium silicate solution**.

■ **water of crystallization** Definite amount of water retained by a compound (usually a salt) when crystallized from solution. The chemical formula of the resulting **hydrate** shows the number of molecules of water of crystallization associated with each molecule of hydrate; *e.g.* $Na_2SO_4.7H_2O$. The water can usually be removed by heating, and the resulting compound is termed **anhydrous**. Alternative name: water of hydration.

wave function Mathematical equation that expresses time and space variation in amplitude for a wave system. *See* **Schrödinger's equation**.

■ **weak acid** Acid that shows little **ionization** or dissociation in solution; *e.g.* **carbonic acid**, **acetic** (ethanoic) **acid**.

white arsenic Alternative name for arsenic(III) oxide (arsenious oxide). *See* **arsenic**.

■ **white spirit** **Petroleum** distillate used as a solvent and in the manufacture of paint and varnish.

white vitriol Alternative name for **zinc sulphate**.

Williamson's synthesis Method of synthesizing **ethers** from alkyl iodides and sodium alcoholates. It was named after the British chemist Alexander Williamson (1824–1904).

will-o'-the-wisp *See* **ignis fatuus**.

Wöhler's synthesis Method of preparing **urea** by heating ammonium isocyanate. It was the first synthesis of an organic compound (at that time, 1828, thought to be produced only by living organisms) from an inorganic one. It was named after the German chemist Friedrich Wöhler (1800–82).

■ **wood spirit** Alternative name for **methanol**.

wood sugar Alternative name for **xylose**.

wool fat Alternative name for **lanolin**.

Wurtz reaction Method of synthesizing **hydrocarbons** using alkyl iodides and sodium metal. It was named after the French chemist Charles-Adolphe Wurtz (1817–84).

X

xanthate **Salt** or **ester** of **xanthic acid**. Sodium and potassium xanthates are used as ore flotation collectors; cellulose xanthate is used in the manufacture of **rayon**.

xanthene $CH_2(C_6H_4)_2O$ Yellow crystalline organic compound, used as a fungicide and in making dyes. Alternative name: tricyclicdibenzopyran.

xanthic acid $CS(OC_2H_5)SH$ Organic acid whose salts and esters (**xanthates**) have various industrial applications.

xanthine $C_5H_4N_2O_2$ Toxic organic compound that occurs in small amounts in potatoes, coffee beans, blood and urine, used industrially as a chemical intermediate. Alternative name: 3,7-dihydro-1H-purine-2,6-dione, 2,6-dihydroxypurine.

xanthone $CO(C_6H_4)_2O$ Plant **pigment** that occurs in gentians and other flowers, used commercially as an insecticide and dye intermediate. Alternative name: 9H-xanthen-9-one.

■ **xenon** Xe Unreactive gaseous element in Group 0 of the Periodic Table (the **rare gases**) which occurs as traces in the atmosphere, from which it is extracted. It is used in electronic flash tubes and high-intensity arc lamps. The **isotope** xenon-135 is a uranium fission product and a troublesome 'poison' in **nuclear reactors** (because it captures slow neutrons). At. no. 54; r.a.m. 131.30. [6/9/a]

X-ray crystallography Study of **crystal** structure by examination of the diffraction pattern obtained when a beam of **X-rays** is passed through the crystal **lattice**.

X-ray diffraction Pattern of variable intensities produced by diffraction of **X-rays** when passed through a diffraction grating consisting of spacings of about 10^{-8}cm, in particular that formed by the **lattice** of a crystal.

X-rays Electromagnetic radiation produced in partial vacuum by the sudden arrest of high-energy bombarding electrons as they collide with the heavy atom nuclei of a target metal. The X-rays produced are thus characteristic of the target's atoms. X-rays have very short wavelengths (10^{-3} to 1 nm) and can penetrate solids to varying degrees; this characteristic has made them useful in medicine, dentistry and X-ray crystallography. Alternative name: röntgen (roentgen) rays; X-radiation.

xylene $C_6H_4(CH_3)_2$ Aromatic liquid organic compound that exists in three isomeric forms (ortho-, meta- and para-xylene), obtained from coal-tar and petroleum. They are used as solvents in **polyester** synthesis, in microscopy for preparation of specimens and as cleaning agents. Alternative name: dimethylbenzene.

xylose $C_5H_{10}O_5$ Naturally occurring **pentose sugar**, found in the form of xylan or as **glycosides** in many plants (*e.g.* cherry and maple wood, straw, pecan shell, corn cobs and cotton-seed hulls). Alternative name: wood sugar.

Y

ytterbium Yb Silvery-white metallic element in Group IIIA of the Periodic Table (one of the **lanthanides**), with no commercial uses. At. no. 70; r.a.m. 173.04.

yttrium Y Grey metallic element in Group IIIA of the Periodic Table (one of the **alkaline earths**, but often classed with the **lanthanides**). It is used in alloys for superconductors and magnets, and yttrium(VI) oxide, Y_2O_6, is employed in lasers and **phosphors**. At. no.39; r.a.m. 88.905.

Z

■ **zeolite** Hydrated aluminosilicate mineral, from which the water is easily removed, used for making **molecular sieves** and **ion exchange** columns (*e.g.* for water softeners).

Ziegler catalyst Catalyst made by mixing certain **transition element** salts (*e.g.* titanium(IV) chloride, $TiCl_4$) with an organometallic compound (*e.g.* triethylaluminium, $Al(C_2H_5)_3$). It is used in the **Ziegler process**. It was named after the German chemist Karl Ziegler (1898–1973).

Ziegler process Commercial process for making high-density **polymers** (*e.g.* polyethylene, stereospecific rubbers) using **Ziegler catalyst**.

■ **zinc** Zn Bluish-white metallic element in Group IIB of the Periodic Table (a **transition element**), used to give a corrosion-resistant coating to steel (**galvanizing**), to make dry batteries and in various alloys (*e.g.* brass, bronze). Its oxide, ZnO, is used as a white pigment (Chinese white). At. no. 30; r.a.m. 65.38.

zincate Compound formed by the reaction of metallic **zinc** or zinc oxide with an alkali; *e.g.* Na_2ZnO_2.

zinc chloride $ZnCl_2$ White **hygroscopic** salt produced commercially by heating metallic **zinc** in dry **chlorine** gas. It is used to fireproof timber, in battery making, vulcanizing and galvanizing, in oil refining, and as a fungicide and catalyst.

zincite Deep red-orange mineral, mainly zinc oxide, which is an important **zinc** ore; it often also contains some manganese. Alternative names: red oxide of zinc, spartalite.

zinc oxide ZnO White crystalline solid (yellow when hot) which can be produced directly by heating zinc in air. It is used as a white pigment (Chinese white), in ceramics, cosmetics, pharmaceuticals and floor coverings, and in the manufacture of tyres. It dissolves in alkalis to form **zincates**.

zinc sulphate $ZnSO_4.7H_2O$ Colourless crystalline salt prepared by dissolving metallic **zinc** in dilute **sulphuric acid**. It is used in manufacture of rayon, glue, fertilizers, fungicides, wood preservatives, rubber, paint and varnishes. Alternative names: white copperas, white vitriol, zinc vitriol.

zinc sulphide ZnS Occurs naturally as blende (an important **zinc** ore) and can be prepared as a white precipitate by adding ammonium sulphide or hydrogen sulphide to a solution of a zinc salt. It is used as the pigmentary base for white zinc sulphide (lithopone), which contains up to 60% zinc sulphide and a balance of **barium sulphate**. It is also used in fungicides and **phosphors**.

zircon $ZrSiO_4$ Mineral form of zirconium silicate, a pale blue, golden-yellow, red or greyish substance which is the chief source of **zirconium**. Colourless crystals are employed as semi-precious gemstones.

zirconium Zr Silvery-grey metallic element in Group IVA of the Periodic Table (a **transition element**). It is used to clad uranium fuel rods in **nuclear reactors**. Naturally occurring crystalline zirconium(IV) oxide, ZrO_2, is the semi-precious gemstone zircon; the oxide is also used as an **electrolyte** in **fuel cells**. At. no. 40; r.a.m. 91.22.

zone *1.* In analytical chemistry, orientation of solute molecules in a series of tubes in a liquid-liquid extraction procedure. *2.* In crystallography, crystal faces that intersect along parallel edges.

■ **zymase** **Enzyme** that catalyses the **fermentation** of **carbohydrates** to **ethanol** (ethyl alcohol).

zymogen Inactive precursor of an **enzyme** formed by plants and animals. It is activated by the action of a kinase. Alternative name: proenzyme.

APPENDIX I

SI Units

Basic unit	*Symbol*	*Quantity*	*Standard*
metre	m	length	Distance light travels in vacuum in 1/299792458 of a second
kilogram	kg	mass	Mass of the international prototype kilogram, a cylinder of platinum-iridium alloy (kept at Sèvres, France)
second	s	time	Time taken for 9,192,631,770 resonance vibrations of an atom of caesium-133
kelvin	K	temperature	1/273.16 of the temperature of the triple point of water
ampere	A	electric current	Current that produces a force of 2×10^{-7} newtons per metre between two parallel conductors of infinite length and negligible cross-section placed a metre apart in vacuum
mole	mol	amount of substance	Amount of substance that contains as many atoms (or molecules, ions, or subatomic

			particles) as 12 grams of carbon-12 has atoms
candela	cd	luminous intensity	Luminous intensity of a source that emits monochromatic light of frequency 540 × 10^{12} hertz of radiant intensity 1/683 watt per steradian in a given direction
Supplementary units			
radian	rad	plane angle	Angle subtended at the centre of a circle by an arc whose length is the radius of the circle
steradian	sr	solid angle	Solid angle subtended at the centre of a sphere by a part of the surface whose area is equal to the square of the radius of the sphere
Derived units			
becquerel	Bq	radioactivity	Activity of a quantity of a radioisotope in which 1 nucleus decays every second (on average)
coulomb	C	electric charge	Charge that is carried by a current of 1 ampere flowing for 1 second
farad	F	electric capacitance	Capacitance that holds a charge of 1 coulomb when it is charged by a potential difference of 1 volt

gray	Gy	absorbed dose	Dosage of ionizing radiation corresponding to 1 joule of energy per kilogram
henry	H	inductance	Mutual inductance in a closed circuit in which an electromotive force of 1 volt is produced by a current that varies at 1 ampere per second
hertz	Hz	frequency	Frequency of 1 cycle per second
joule	J	energy	Work done when a force of 1 newton moves its point of application 1 metre in its direction of application
lumen	lm	luminous flux	Amount of light emitted per unit solid angle by a source of 1 candela intensity
lux	lx	illuminance	Amount of light that illuminates 1 square metre with a flux of 1 lumen
newton	N	force	Force that gives a mass of 1 kilogram an acceleration of 1 metre per second squared
ohm	Ω	electric resistance	Resistance of a conductor across which a potential of 1 volt produces a current of 1 ampere

pascal	Pa	pressure	Pressure exerted when a force of 1 newton acts on an area of 1 square metre
siemens	S	electric conductance	Conductance of a material or circuit component that has a resistance of 1 ohm
sievert	Sv	dose equivalent	Radiation dosage equal to 1 joule of radiant energy per kilogram
tesla	T	magnetic flux density	Flux density (or magnetic induction) of 1 weber of magnetic flux per square metre
volt	V	electric potential difference	Potential difference across a conductor in which a constant current of 1 ampere dissipates 1 watt of power
watt	W	power	Amount of power equal to a rate of energy transfer of (or of doing work at) 1 joule per second
weber	Wb	magnetic flux	Amount of magnetic flux that, decaying to zero in 1 second, induces an electromotive force of 1 volt in a circuit of one turn

APPENDIX II

Chemical Elements

Name	*Symbol*	*At. no.*	*R.a.m.*
actinium	Ac	89	(227)
aluminium	Al	13	26.9815
americium	Am	95	(243)
antimony	Sb	51	121.75
argon	Ar	18	39.948
arsenic	As	33	74.9216
astatine	At	85	(210)
barium	Ba	56	137.34
berkelium	Bk	97	(247)
beryllium	Be	4	9.0122
bismuth	Bi	83	208.9806
boron	B	5	10.81
bromine	Br	35	79.904
cadmium	Cd	48	112.40
caesium	Cs	55	132.9055
calcium	Ca	20	40.08
californium	Cf	98	(251)
carbon	C	6	12.001
cerium	Ce	58	140.12
chlorine	Cl	17	35.453
chromium	Cr	24	51.996
cobalt	Co	27	58.9332
copper	Cu	29	63.546
curium	Cm	96	(247)
dysprosium	Dy	66	162.50
einsteinium	Es	99	(254)
erbium	Er	68	167.26

europium	Eu	63	151.96
fermium	Fm	100	(257)
fluorine	F	9	18.9984
francium	Fr	87	(223)
gadolinium	Gd	64	157.25
gallium	Ga	31	69.72
germanium	Ge	32	72.59
gold	Au	79	196.9665
hafnium	Hf	72	178.49
hahnium	Ha	105	–
helium	He	2	4.0026
holmium	Ho	67	164.9303
hydrogen	H	1	1.0080
indium	In	49	114.82
iodine	I	53	126.904
iridium	Ir	77	192.22
iron	Fe	26	55.847
krypton	Kr	36	83.80
lanthanum	La	57	138.9055
lawrencium	Lr	103	(257)
lead	Pb	82	207.19
lithium	Li	3	6.941
lutetium	Lu	71	174.97
magnesium	Mg	12	24.305
manganese	Mn	25	54.9380
mendelevium	Md	101	(258)
mercury	Hg	80	200.59
molybdenum	Mo	42	95.94
neodymium	Nd	60	144.24
neon	Ne	10	20.179
neptunium	Np	93	(237)
nickel	Ni	28	58.71
niobium	Nb	41	92.9064
nitrogen	N	7	14.0067

nobelium	No	102	(255)
osmium	Os	76	190.2
oxygen	O	8	15.9994
palladium	Pd	46	106.4
phosphorus	P	15	30.9738
platinum	Pt	78	195.09
plutonium	Pu	94	(244)
polonium	Po	84	(209)
potassium	K	19	39.102
praeseodymium	Pr	59	140.9077
promethium	Pm	61	(145)
protactinium	Pa	91	231.0359
radium	Ra	88	226.0254
radon	Rn	86	(222)
rhenium	Re	75	186.20
rhodium	Rh	45	102.9055
rubidium	Rb	37	85.4678
ruthenium	Ru	44	101.07
rutherfordium	Rf	104	–
samarium	Sm	62	150.35
scandium	Sc	21	44.9559
selenium	Se	34	78.96
silicon	Si	14	28.086
silver	Ag	47	107.868
sodium	Na	11	22.9898
strontium	Sr	38	87.62
sulphur	S	16	32.06
tantalum	Ta	73	180.9479
technetium	Tc	43	(99)
tellurium	Te	52	127.60
terbium	Tb	65	158.9254
thallium	Tl	81	204.39
thorium	Th	90	232.0381
thulium	Tm	69	168.9342

tin	Sn	50	118.69
titanium	Ti	22	47.90
tungsten	W	74	183.85
uranium	U	92	238.029
vanadium	V	23	50.9414
xenon	Xe	54	131.30
ytterbium	Yb	70	173.04
yttrium	Y	39	88.9059
zinc	Zn	30	65.38
zirconium	Zr	40	91.22

Relative atomic masses in brackets are those of the longest-lived isotopes.

APPENDIX III

Nobel Prizewinners in Chemistry

1901	J. van't Hoff (Dutch): laws of chemical dynamics and osmotic pressure
1902	E. Fischer (German): organic syntheses
1903	S. Arrhenius (Swedish): theory of ionic dissociation
1904	W. Ramsay (British): discovery of helium, krypton, neon and xenon
1905	A. von Baeyer (German): synthesis of indigo etc.
1906	H. Moissan (French): discovery of fluorine and development of electric furnace
1907	E. Buchner (German): research in biochemistry
1908	E. Rutherford (New Zealand/British): study of alpha particles and radioactivity
1909	W. Ostwald (German): study of chemical reactions and catalysis
1910	O. Wallach (German): study of alicyclic compounds
1911	M. Curie (French): discovery of radium and polonium
1912	V. Grignard and P. Sabatier (French): organic syntheses
1913	A. Werner (Swiss): study of co-ordination compounds
1914	T. Richards (American): atomic weight determinations
1915	R. Willstätter (German): study of chlorophyll
1916–17	*No award*
1918	F. Haber (German): industrial synthesis of ammonia
1919	*No award*

1920	W. Nernst (German): study of heats of reaction
1921	F. Soddy (British): study of isotopes
1922	F. Aston (British): mass spectroscopy
1923	F. Pregl (Austrian): organic microanalysis
1924	*No award*
1925	R. Zsigmondy (German): study of colloids
1926	T. Svedberg (Swedish): study of colloids and dispersions
1927	H. Wieland (German): study of bile acids etc.
1928	A. Windaus (German): study of steroids
1929	A. Harden (British) and H. von Euler-Chelpin (German/Swedish): study of enzymes and fermentation
1930	H. Fischer (German): study of haemoglobin and chlorophyll
1931	C. Bosch and F. Bergius (German): ammonia synthesis and hydrogenation of coal
1932	I. Langmuir (American): chemistry of adsorbed layers on surfaces
1933	*No award*
1934	H. Urey (American): discovery of deuterium
1935	F. and I. Joliot-Curie (French): synthesis of radioactive isotopes
1936	P. Debye (Dutch): study of dipole moments, electron diffraction and X-rays in gases
1937	W. Haworth (British) and P. Karrer (Swiss): studies of vitamins
1938	R. Kühn (German): research on vitamins (Kühn forced to decline; award presented in 1946)
1939	A. Butenandt (German) and L. Ruzicka (Swiss): studies of sex hormones and polymethylenes (Butenandt forced to decline)
1940–42	*No award*

1943	G. von Hevesy (Hungarian/Swedish): isotopic tracers in chemistry
1944	O. Hahn (German): discoveries in nuclear fission
1945	A. Virtanen (Finnish): studies in agricultural biochemistry
1946	J. Northrop, W. Stanley and J. Sumner (American): preparation of pure enzymes
1947	R. Robinson (British): plant biochemistry
1948	A. Tiselius (Swedish): study of serum proteins
1949	W. Giauque (American): cryogenics
1950	O. Diels and K. Alder (German): organic syntheses
1951	E. McMillan and G. Seaborg (American): discovery of plutonium etc.
1952	A. Martin and R. Synge (British): partition chromatography
1953	H. Staudinger (German): theory of macromolecular chains
1954	L. Pauling (American): theories on interatomic forces
1955	V. Du Vigneaud (American): hormone synthesis
1956	C. Hinshelwood (British) and N. Semenov (Soviet): studies of chemical chain reactions
1957	Lord Todd (British): composition of cell proteins
1958	F. Sanger (British): structure of insulin
1959	J. Heyrovsky (Czech): development of polarography
1960	W. Libby (American): radio-carbon dating
1961	M. Calvin (American): study of photosynthesis
1962	J. Kendrew and M. Perutz (British): structures of globular proteins
1963	G. Natta (Italian) and K. Ziegler (German): study of polymers and polymerization reactions
1964	D. Hodgkin (British): X-ray analysis of large organic molecules

1965	R. Woodward (American): organic syntheses
1966	R. Mulliken (American): molecular orbital theory
1967	M. Eigen (German), R. Norrish and G. Porter (British): measurement of rates of reaction
1968	L. Onsager (American): study of non-equilibrium thermodynamics
1969	D. Barton (British) and O. Hassel (Norwegian): effect of stereochemistry on reaction rates
1970	L. Leloir (Argentinian): study of energy-storing biochemicals
1971	G. Herzberg (Canadian): study of free radicals
1972	C. Anfinsén, S. Moore and W. Stein (American): study of emzymes
1973	E. Fischer (W. German) and G. Wilkinson (British): study of organometallic compounds
1974	P. Flory (American): polymer chemistry
1975	J. Warcup (Australian) and V. Prelog (Swiss): organic syntheses
1976	W. Lipscomb Jr (American): structure and chemistry of boranes
1977	I. Prigogine (Belgian): study of non-equilibrium thermodynamics
1978	P. Mitchell (British): study of energy transfer in cells
1979	H. Brown (American) and G. Wittig (German): preparation of organoboron compounds etc. useful in synthesis
1980	P. Berg, W. Gilbert (American) and F. Sanger (British): structure of nucleic acids
1981	K. Fukui (Japanese) and R. Hoffmann (American): application of quantum mechanics to reaction kinetics
1982	A. Klug (South African/British): structure of nucleic acid-protein complexes

1983	H. Taube (American): study of electron transfer in chemical reactions
1984	B. Merrifield (American): development of automated synthesis of peptides
1985	H. Hauptman and J. Karle (American): rapid method of determining structures of biochemical molecules
1986	D. Herschbach, Y. Lee (American) and J. Polanyi (Canadian): research on basic reactions
1987	D. Cram, C. Pedersen (American) and J. Lehn (French): synthesis of biomolecules that behave like natural ones
1988	J.Deisendorfer, R. Huber and H. Michel (German): research into photosynthesis
1989	T. Cech and S. Altman (American): discovery of the catalytic action of DNA
1990	E.J. Corey (American): new methods of chemical synthesis

APPENDIX IV

Periodic Table

IA	IIA	IIIA	IVA	VA	VIA	VIIA	VIII	
1 H								
3 Li	4 Be							
11 Na	12 Mg							
19 K	20 Ca	21 Sc	22 Ti	23 V	24 Cr	25 Mn	26 Fe	27 Co
37 Rb	38 Sr	39 Y	40 Zr	41 Nb	42 Mo	43 Tc	44 Ru	45 Rh
55 Cs	56 Ba	57 La	72 Hf	73 Ta	74 W	75 Re	76 Os	77 Ir
87 Fr	88 Ra	89 Ac						
				58 Ce	59 Pr	60 Nd	61 Pm	62 Sm
				90 Th	91 Pa	92 U	93 Np	94 Pu

	IB	IIB	IIIB	IVB	VB	VIB	VIIB	0
								2 He
			5 B	6 C	7 N	8 O	9 F	10 Ne
			13 Al	14 Si	15 P	16 S	17 Cl	18 Ar
28 Ni	29 Cu	30 Zn	31 Ga	32 Ge	33 As	34 Se	35 Br	36 Kr
46 Pd	47 Ag	48 Cd	49 In	50 Sn	51 Sb	52 Te	53 I	54 Xe
78 Pt	79 Au	80 Hg	81 Ti	82 Pb	83 Bi	84 Po	85 At	86 Rn

63 Eu	64 Gd	65 Tb	66 Dy	67 Ho	68 Er	69 Tm	70 Yb	71 Lu
95 Am	96 Cm	97 Bk	98 Cf	99 Es	100 Fm	101 Md	102 No	103 Lr